Solid Geometry with MATLAB Programming

RIVER PUBLISHERS SERIES IN MATHEMATICAL, STATISTICAL AND COMPUTATIONAL MODELLING FOR ENGINEERING

Series Editors:

MANGEY RAM
Graphic Era University, India

TADASHI DOHI
Hiroshima University, Japan

ALIAKBAR MONTAZER HAGHIGHI
Prairie View Texas A&M University, USA

Applied mathematical techniques along with statistical and computational data analysis has become vital skills across the physical sciences. The purpose of this book series is to present novel applications of numerical and computational modelling and data analysis across the applied sciences. We encourage applied mathematicians, statisticians, data scientists and computing engineers working in a comprehensive range of research fields to showcase different techniques and skills, such as differential equations, finite element method, algorithms, discrete mathematics, numerical simulation, machine learning, probability and statistics, fuzzy theory, etc.

Books published in the series include professional research monographs, edited volumes, conference proceedings, handbooks and textbooks, which provide new insights for researchers, specialists in industry, and graduate students.

Topics included in this series are as follows:-

- Discrete mathematics and computation
- Fault diagnosis and fault tolerance
- Finite element method (FEM) modeling/simulation
- Fuzzy and possibility theory
- Fuzzy logic and neuro-fuzzy systems for relevant engineering applications
- Game Theory
- Mathematical concepts and applications
- Modelling in engineering applications
- Numerical simulations
- Optimization and algorithms
- Queueing systems
- Resilience
- Stochastic modelling and statistical inference
- Stochastic Processes
- Structural Mechanics
- Theoretical and applied mechanics

For a list of other books in this series, visit www.riverpublishers.com

Solid Geometry with MATLAB Programming

Nita H. Shah
Gujarat University, India

Falguni S. Acharya
Parul University, India

NEW YORK AND LONDON

Published 2023 by River Publishers
River Publishers
Alsbjergvej 10, 9260 Gistrup, Denmark
www.riverpublishers.com

Distributed exclusively by Routledge
605 Third Avenue, New York, NY 10017, USA
4 Park Square, Milton Park, Abingdon, Oxon OX14 4RN

Solid Geometry with MATLAB Programming / by Nita H. Shah, Falguni S. Acharya.

© 2023 River Publishers. All rights reserved. No part of this publication may be reproduced, stored in a retrieval systems, or transmitted in any form or by any means, mechanical, photocopying, recording or otherwise, without prior written permission of the publishers.

Routledge is an imprint of the Taylor & Francis Group, an informa business

ISBN 978-87-7022-761-2 (print)
ISBN 978-10-0082-412-4 (online)
ISBN 978-10-0336-068-1 (ebook master)

While every effort is made to provide dependable information, the publisher, authors, and editors cannot be held responsible for any errors or omissions.

Contents

Preface ix

1 Plane 1
1.1 Definition 1
1.2 General Equation of the First Degree in x, y, z Represents a Plane 1
1.3 Transformation of General form to Normal Form 3
1.4 Direction Cosines of the Normal to a Plane 4
1.5 Equation of a Plane Passing through a Given Point ... 5
1.6 Equation of the Plane in Intercept Form 6
1.7 Reduction of the General Equation of the Plane to the Intercept Form 7
1.8 Equation of a Plane Passing through three Points 10
1.9 Equation of any Plane Parallel to a Given Plane 15
1.10 Equation of Plane Passing through the Intersection of Two Given Planes 16
1.11 Equation of the Plane Passing through the Intersection ... 17
1.12 Angle between Two Planes 21
1.13 Position of the Origin w.r.t. the Angle between Two Planes . 23
1.14 Two Sides of a Plane 24
1.15 Length of the Perpendicular from a Point to a Plane 26
1.16 Bisectors of Angles between Two Planes 28
1.17 Pair of Planes 31
1.18 Orthogonal Projection on a Plane 35
1.19 Volume of a Tetrahedron 36
    Exercise 42

2 Straight Line 45
2.1 Representation of Line (Introduction) 45
2.2 Equation of a Straight Line in the Symmetrical Form .. 45

2.3	Equation of a Straight Line Passing through Two Points	46
2.4	Transformation from the Unsymmetrical to the Symmetrical Form	49
2.5	Angle between a Line and a Plane	53
2.6	Point of Intersection of a Line and a Plane	54
2.7	Conditions for a Line to Lie in a Plane	55
2.8	Condition of Coplanarity of Two Straight Lines	56
2.9	Skew Lines and the Shortest Distance between Two Lines	69
2.10	Equation of Two Skew Lines in Symmetric Form	72
2.11	Intersection of Three Planes	80
	Exercise	87

3 Sphere — 89

3.1	Definition	89
3.2	Equation of Sphere in Vector Form	89
3.3	General Equation of the Sphere	91
3.4	Equation of Sphere Whose End-Points of a Diameter are Given	91
3.5	Equation of a Sphere Passing through the Four Points	93
3.6	Section of the Sphere by a Plane	105
3.7	Intersection of Two Spheres	106
3.8	Intersection of Sphere S and Line L	115
3.9	Tangent Plane	116
3.10	Equation of the Normal to the Sphere	117
3.11	Orthogonal Sphere	127
	Exercise	129

4 Cone — 133

4.1	Definition	133
4.2	Equation of a Cone with a Conic as Guiding Curve	133
4.3	Enveloping Cone to a Surface	138
4.4	Equation of the Cone whose Vertex is the Origin is Homogeneous	142
4.5	Intersection of a Line with a Cone	149
4.6	Equation of a Tangent Plane at (α, β, γ) to the Cone with Vertex Origin	150
4.7	Conditions for Tangency	152
4.8	Right Circular Cone	156
	Exercise	164

5	**Cylinder**	**167**
	5.1 Definition	167
	5.2 Equation of the Cylinder whose Generators Intersect the Given Conic	168
	5.3 Enveloping Cylinder	170
	5.4 Right Circular Cylinder	175
	Exercise	182
6	**Central Conicoid**	**185**
	6.1 Definition	185
	6.2 Intersection of a Line with the Central Conicoid	185
	6.3 Tangent Lines and Tangent Plane at a Point	186
	6.4 Condition of Tangency	188
	6.5 Normal to Central Conicoid	191
	6.6 Plane of Contact	197
	6.7 Polar Plane of a Point	197
	Exercise	201
7	**Miscellaneous Examples using MATLAB**	**203**
Index		**227**
About the Authors		**229**

Preface

Solid geometry is applied to analyze, predict and calculate the volume, location, and surface area of all the objects in the three-dimensional space. Its fundamental use is for defining three-dimensional real-life objects.

The authors have designed the book following the significance of the subject in day-to-day life and the challenges students face to learn Solid Geometry. The book is formulated not only for students of bachelor level but also for all levels coherently with ample examples. The content includes both generic and advanced topics of Solid Geometry.

It is a handy reference book and will serve as a fundamental resource for advanced-level studies. The book can aid learning's in various fields such as engineering, interior designing, and architecture.

Unlike many competitive books in the market, this book involves MATLAB programming. Authors have avoided unnecessary information and diverting topics, providing a lucid and informative explanation for theoretical aspects. The book has seven chapters, out of which the first six chapters include basics, applications, and advanced level of topics. The seventh chapter focuses entirely on programming through MATLAB, which gives this book a touch of uniqueness.

The author is hopeful that this book will suffice the requirement of elementary knowledge to students and learners from diverse genres. It is a one-stop platform to understand solid geometry for beginners. The author welcomes suggestions for improvements and observed errors.

1

Plane

1.1 Definition

A locus is a plane such that if any two points P and Q are on the locus then every point of the line PQ lies on the locus.

1.2 General Equation of the First Degree in x, y, z Represents a Plane

Let the first-degree equation of the plane in x, y, z be $ax + by + cz + d = 0$, where a, b and c are given nonzero real numbers i.e., $a^2 + b^2 + c^2 \neq 0$.

Theorem: Every equation of the first degree in x, y, z represents the a plane.

Proof: Let us consider the first-degree equation in x, y, z.

$$ax + by + cz + d, a^2 + b^2 + c^2 \neq 0. \tag{1.1}$$

Let $P(x_1, y_1, z_1)$, and $Q(x_2, y_2, z_2)$ be two points on the locus, so we get

$$ax_1 + by_1 + cz_1 + d = 0, \tag{1.2}$$

$$ax_2 + by_2 + cz_2 + d = 0. \tag{1.3}$$

Multiplying Equation (1.2) by m and adding to Equation (1.1), we get

$$a(x_1 + mx_2) + b(y_1 + my_2) + c(z_1 + mz_2) + d(1 + m) = 0,$$

$$\therefore a\left(\frac{x_1+mx_2}{1+m}\right) + b\left(\frac{y_1+my_2}{1+m}\right) + c\left(\frac{z_1+mz_2}{1+m}\right) + d = 0.$$

Suppose $m + 1 \neq 0$. i.e., $m \neq -1$, which clearly shows that the point

$$\left(\frac{x_1 + mx_2}{1 + m}, \frac{y_1 + my_2}{1 + m}, \frac{z_1 + mz_2}{1 + m}\right),$$

2 Plane

is a point on the locus for every value of $m \neq -1$.

Thus, we have proved that if $P(x_1, y_1, z_1)$, and $Q(x_2, y_2, z_2)$ lies on (1.1) then every point R of the line joining P and Q lies on (1.1).

i.e., Every point on the straight line joining any two arbitrary points on the locus also lies on the locus.

∴ The given Equation (1.1) represents a plane.

i.e., The general equation of the first degree in x, y, z represents a plane.

Converse: (Normal form)

The equation of every plane is of the first degree.

i.e., of form $ax + by + cz + d = 0$ where $a^2 + b^2 + c^2 \neq 0$.

Proof: Let us consider any plane and P be the length of the perpendicular from the origin to the plane and l, m, n be the direction cosines of this perpendicular.

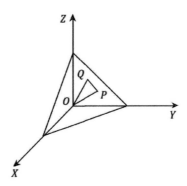

We shall prove that for any point (x, y, z) on the plane, we have the relation

$$lx + my + nz = p.$$

i.e., The equation of the plane is of the first degree.

Let Q be the foot of the perpendicular from the origin O to the plane.
Let $OQ = p$ and l, m, n be its direction cosines.
Let $P(x, y, z)$ be any point on the plane.
∴ PQ lies in the plane.

$$\Rightarrow PQ \perp OQ$$

$\Rightarrow$ The projections of OP on $OQ = OQ = p$.

Also, the projection of the segment OP joining the points $O(0,0,0)$ and $P(x,y,z)$ on the line OQ with direction cosines l, m, n is

$$l(x-0) + m(y-0) + n(z-0) = lx + my + nz,$$

i.e., $lx + my + nz = p$ which satisfies the coordinates of any point $P(x,y,z)$ on the given plane.

∴ This equation is the equation of the plane.

Remark:

1) An equation $lx + my + nz = p$ is in the normal form if

    1.1 (coefficient of x)^2 + (coefficient of y)^2 + (coefficient of z)^2 = 1
    i.e., $l^2 + m^2 + n^2 = 1$.

    1.2 constant term on the R.H.S. is positive, i.e., $p > 0$.

2) If the plane passes through the origin (0,0,0) then $d = 0$ and equation of a plane passing through the origin is $ax + by + cz + d = 0$.
3) Equation of any plane parallel to the plane $ax + by + cz + d = 0$ is given by $ax + by + cz + d_1 = 0$.

1.3 Transformation of General form to Normal Form

To transform the equation

$$ax + by + cz + d = 0; a^2 + b^2 + c^2 \neq 0, \qquad (1.4)$$

to the normal form

$$lx + my + nz = p. \qquad (1.5)$$

Equations (1.4) and (1.5) represent the same plane so comparing the coefficient we get

$$\frac{a}{l} = \frac{b}{m} = \frac{c}{n} = \frac{-d}{p},$$

$$\therefore l = \frac{-pa}{d}; m = \frac{-pb}{d}; n = \frac{-pc}{d}.$$

$l, m,$ and n are direction cosines of the normal to the plane

$$\therefore l^2 + m^2 + n^2 = 1$$
$$\therefore p^2 a^2 + p^2 b^2 + p^2 c^2 = d^2$$
$$\therefore p^2 = \frac{d^2}{a^2 + b^2 + c^2}$$
$$\therefore p = \pm \frac{d}{\sqrt{a^2 + b^2 + c^2}}.$$

Case (i): If d is positive then p is also positive

i.e., $p = \dfrac{d}{\sqrt{a^2+b^2+c^2}}$,

$\therefore l = \dfrac{-a}{\sqrt{a^2+b^2+c^2}}; m = \dfrac{-b}{\sqrt{a^2+b^2+c^2}}$ and $n = \dfrac{-c}{\sqrt{a^2+b^2+c^2}}$.

Substituting the values of $l, m,$ and n in (1.5), we get $ax+by+cz = d$.

Case (ii): If d is negative then p is always positive

$\therefore p = \dfrac{-d}{\sqrt{a^2+b^2+c^2}}$,

$\therefore l = \dfrac{a}{\sqrt{a^2+b^2+c^2}}; m = \dfrac{b}{\sqrt{a^2+b^2+c^2}}$ and $n = \dfrac{c}{\sqrt{a^2+b^2+c^2}}$.

Substituting the values of l, m and n in (1.4), we get $ax+by+cz+d = 0$.

Remark:

To reduce general form to normal form

1) Shift the constant term to R.H.S. and make it positive (if it is not)

2) Divide throughout by $\sqrt{a^2+b^2+c^2}$.

1.4 Direction Cosines of the Normal to a Plane

The direction cosines of the normal to a plane are proportional to the coefficients of x, y, z in the equation.

$\therefore a, b, c$ are direction ratios of the normal to the plane $ax+by+cz+d = 0$.

1) If the axes are rectangular and P is the point $(2, 3, -1)$ find the equation to the plane through P at right angles to OP.

Sol. Given $O(0, 0, 0)$ and $P(2, 3, -1)$

$\therefore$ Direction ratios of OP are $2-0, 3-0, -1-0$.

i.e., Direction ratios of OP are $2, 3, -1$.

$\therefore$ The equation of the plane is $2x + 3y - z + d = 0$.

Since the plane passes through point $P(2, 3, -1)$, we get

$$2(2) + 3(3) - (-1) + d = 0$$
$$\therefore d = -14.$$

$\therefore$ The required equation of the plane is $2x + 3y - z - 14 = 0$.

1.5 Equation of a Plane Passing through a Given Point

Let the equation of a plane in general form be

$$ax + by + cz + d = 0, \tag{1.6}$$

and let it passes through the point (x_1, y_1, z_1).
i.e., Equation (1.6) will become

$$ax_1 + by_1 + cz_1 + d = 0. \tag{1.7}$$

Subtracting (1.7) from (1.6), we get

$$a(x_1 - x) + b(y_1 - y) + c(z_1 - z) = 0.$$

2) Find the ratios in which the coordinate planes divide the line joining $(-2, 4, 7)$ and $(3, -5, 8)$.

Sol. Let R be the point which divides $P(-2, 4, 7)$ and $Q(3, -5, 8)$ in the ratio $m = 1$.
$\therefore$ Coordinates of R are $\left(\frac{3m-2}{m+1}, \frac{-5m+4}{m+1}, \frac{8m+7}{m+1}\right)$.
(a) If R lies on the XY plane i.e., $z = 0$, we have $8m + 7 = 0 \Rightarrow m = \frac{-7}{8}$.
$\therefore XY$ plane divides PQ externally in the ratio $7 : 8$.
(b) If R lies on the YZ plane i.e., $x = 0$, we have $\frac{3m-2}{m+1} = 0 \Rightarrow m = \frac{2}{3}$.
$\therefore YZ$ plane divides PQ internally in the ratio 2:3.
(c) If R lies on the ZX plane i.e., $y = 0$, we have $\frac{-5m+4}{m+1} = 0 \Rightarrow m = \frac{4}{5}$.
$\therefore ZX$ plane divides PQ internally in the ratio 4:5.

3) Find the equation of the plane through the points $(1, -2, 4)$ and $(3, -4, 5)$ and parallel to the x axis.

Sol. Let the equation of the plane parallel to the x axis be

$$by + cz + d = 0. \tag{1.8}$$

Since it passes through the points $(1, -2, 4)$ and $(3, -4, 5)$, we get

$$-2b + 4z + d = 0, \tag{1.9}$$

$$-4b + 5z + d = 0. \tag{1.10}$$

Solving Equations (1.9) and (1.10), we get $b = \frac{-d}{6}$ and $c = \frac{-2d}{6}$
Substituting the values of b and c in Equation (1.9), we get $\frac{-d}{6}y \frac{-dz}{6} + d = 0$.

$\therefore y + 2z - 6 = 0$ is the required equation of the plane.

4) Find the equation of the plane that passes through $(2, -3, 1)$ and is perpendicular to the line joining the points $(3, -4, 1)$ and $(2, -1, 5)$.

Sol. The equation of the plane passing through the point $(2, -3, 1)$ is

$$a(x - 2) + b(y + 3) + c(z - 1) = 0 \tag{1.11}$$

The direction ratios of the line joining points $P(3, 4, -1)$ and $Q(2, -1, 5)$ are $x_1 - x_2, y_1 - y_2, z_1 - z_2 = 1, 5, -6$.
The required equation of the plane is perpendicular to the line PQ.
$\therefore$ The normal to the plane (1.11) whose direction ratios are a, b, c is parallel to PQ whose direction ratios are $1, 5, -6$.

$$\therefore \frac{a}{1} = \frac{b}{5} = \frac{c}{-6} = t.$$

Substituting $a, b,$ and c in Equation (1.11), we get

$$t(x - 2) + 5t(y + 3) - 6t(x - 1) = 0$$
$$\therefore x - 2 + 5y + 15 - 6z + 6 = 0$$
$$\therefore x + 5y + 6z + 19 = 0$$

is the required equation of the plane.

1.6 Equation of the Plane in Intercept Form

Let ABC be a plane intersecting the coordinate axes OX, OY, OZ at $A, B,$ and C respectively, and let $OA = a, OB = b,$ and $OC = c$.

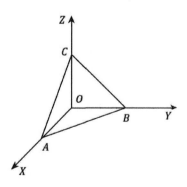

1.7 Reduction of the General Equation of the Plane to the Intercept Form

Then the coordinates of A, B, C are $(a,0,0)$, $(0,b,0)$ and $(0,0,c)$ respectively.

Let the equation of the plane be

$$Px + Qy + Rz + S = 0. \qquad (1.12)$$

Since it passes through point $A(a,0,0)$, we get

$$P(a) + Q(0) + R(0) + S = 0$$
$$\therefore P = \frac{-S}{a}.$$

Similarly, the plane passes through points $C(0,b,0)$ and $D(0,0,c)$ so

$$Q = \frac{-S}{b} \text{ and } R = \frac{-S}{c}$$

Substituting values of P, Q, and R in Equation (1.12), we get

$$\frac{-S}{a}x - \frac{S}{b}y - \frac{S}{c}z + S = 0$$
$$\therefore \frac{x}{a} + \frac{y}{b} + \frac{z}{c} = 1$$

is the required equation of the plane in the intercept form.

Remark:
If the plane passes through the origin, then $S = 0$, and the equation cannot be expressed in intercept form.

1.7 Reduction of the General Equation of the Plane to the Intercept Form

Let the general equation of the plane by

$$ax + by + cz + d = 0 \qquad (1.13)$$

Dividing Equation (1.13) by $(-d)$, we get

$$\frac{ax}{-d} + \frac{by}{-d} + \frac{cz}{-d} = 1,$$
$$\therefore \frac{x}{\frac{-d}{a}} + \frac{y}{\frac{-d}{b}} + \frac{z}{\frac{-d}{c}} = 1.$$

which is of the form

$$\frac{x}{A} + \frac{y}{B} + \frac{z}{C} = 1, \qquad (1.14)$$

8 Plane

where $A = \frac{-d}{a}$, $B = \frac{-d}{b}$, and $C = \frac{-d}{c}$ are the lengths of the intercepts of (1.13) on the axes respectively.

The Equation (1.14) is the intercept form of the general Equation (1.13).

5) A variable plane moves so that the sum of reciprocals of its intercepts on the three co-ordinates axes is constant. Show that it passes through a fixed point.

Sol. Let the variable plane be

$$\frac{x}{a} + \frac{y}{b} + \frac{z}{c} = 1. \qquad (1.15)$$

$\therefore a, b,$ and c are the intercepts on the axes.
Given $\frac{1}{a} + \frac{1}{b} + \frac{1}{c} = $ constant $= \frac{1}{t}$. (Suppose)

$$\therefore \frac{t}{a} + \frac{t}{b} + \frac{t}{c} = 1. \qquad (1.16)$$

$\therefore$ The Equation (1.16) shows that (t, t, t) satisfies Equation (1.15) of the plane.

i.e., plane (1.15) passes through the fixed point (t, t, t).

6) A plane meets the coordinate axes at A, B, C such that the centroid of the triangle ABC is the point (a, b, c). Show that the equation of the plane is

$$\frac{x}{a} + \frac{y}{b} + \frac{z}{c} = 3.$$

Sol. Let the equation of the plane be

$$\frac{x}{r} + \frac{y}{s} + \frac{z}{t} = 1. \qquad (1.17)$$

It meets the axes in $A(r, 0, 0)$, $B(0, s, 0)$, and $C(0, 0, t)$.
Let the centroid G of the $\triangle ABC$ be $\left(\frac{r+0+0}{3}, \frac{0+s+0}{3}, \frac{0+0+t}{3}\right) = \left(\frac{r}{3}, \frac{s}{3}, \frac{t}{3}\right)$.
But it is given that the centroid of $\triangle ABC$ is (a, b, c).

$$\therefore \frac{r}{3} = a \Rightarrow r = 3a, \quad \frac{s}{3} = b \Rightarrow s = 3b, \quad \frac{t}{s} = c \Rightarrow t = 3c$$

Substituting the values of r, s, t in (1.17), we get

$$\frac{x}{3a} + \frac{y}{3b} + \frac{z}{3c} = 1$$

1.7 Reduction of the General Equation of the Plane to the Intercept Form

$$\therefore \frac{x}{a} + \frac{y}{b} + \frac{z}{c} = 3$$

is the required equation of the plane.

7) A variable plane is at a constant distance p from the origin and meets the axes in A, B, C. Through A, B, C planes are drawn parallel to the coordinate planes. Prove that the locus of their point of intersection is $\frac{1}{x^2} + \frac{1}{y^2} + \frac{1}{z^2} = \frac{1}{p^2}$.

Sol. Let the variable plane at distance p from the origin be

$$lx + ny + nz = p, \tag{1.18}$$

where l, m, n are the direction cosines of the normal to the plane.

The plane (1.18) meets the x axis (i.e., $y = 0, z = 0$), we get $lx = p \Rightarrow x = \frac{p}{l}$

$\therefore$ The coordinates of A are $\left(\frac{p}{l}, 0, 0\right)$.

Similarly, coordinates of B and C on the y axis and z axis are $\left(0, \frac{p}{m}, 0\right)$ and $\left(0, 0, \frac{p}{n}\right)$ respectively.

$\therefore$ The equation of the plane through $A\left(\frac{p}{l}, 0, 0\right)$, and parallel to the YZ plane is $x = \frac{p}{l}$.

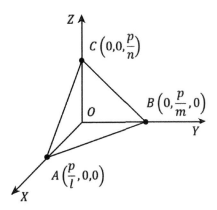

Similarly, the equations of the plane through $B\left(0, \frac{p}{m}, 0\right)$ and $C\left(0, 0, \frac{p}{n}\right)$ parallel to ZX and XY planes respectively are $y = \frac{p}{m}$ and $z = \frac{p}{n}$.

These three planes meet at the point

$$P\left(x = \frac{p}{l}, y = \frac{p}{m}, z = \frac{p}{n}\right)$$

$$\therefore l = \frac{p}{x}; m = \frac{p}{y}; n = \frac{p}{z}.$$

We know that $l^2 + m^2 + n^2 = 1$ ($\because l, m,$ and n are direction cosines)

$$\therefore \frac{p^2}{x^2} + \frac{p^2}{y^2} + \frac{p^2}{z^2} = 1$$

$$\therefore \frac{1}{x^2} + \frac{1}{y^2} + \frac{1}{z^2} = \frac{1}{p^2}$$

1.8 Equation of a Plane Passing through three Points (x_1, y_1, z_1), (x_2, y_2, z_2), and (x_3, y_3, z_3)

Let the equation of the plane be

$$ax + by + cz + d = 0. \tag{1.19}$$

Since it passes through (x_1, y_1, z_1), (x_2, y_2, z_2), and (x_3, y_3, z_3), we get

$$ax_1 + by_1 + cz_1 + d = 0, \tag{1.20}$$

$$ax_2 + by_2 + cz_2 + d = 0, \tag{1.21}$$

$$ax_3 + by_3 + cz_3 + d = 0. \tag{1.22}$$

Eliminating $a, b, c,$ and d from (1.19), (1.20), (1.21), and (1.22), we get

$$\begin{vmatrix} x & y & z & 1 \\ x_1 & y_1 & z_1 & 1 \\ x_2 & y_2 & z_2 & 1 \\ x_3 & y_3 & z_3 & 1 \end{vmatrix} = 0$$

which is the required equation of the plane.

Corollary:

Four points (x_1, y_1, z_1), (x_2, y_2, z_2), (x_3, y_3, z_3) and (x_4, y_4, z_4) will be coplanar if

$$\begin{vmatrix} x_1 & y_1 & z_1 & 1 \\ x_2 & y_2 & z_2 & 1 \\ x_3 & y_3 & z_3 & 1 \\ x_4 & y_4 & z_4 & 1 \end{vmatrix} = 0.$$

Remark:

1) The equation of the xy plane is $z = 0$. The equation of the yz plane is $x = 0$ and the equation of the xz plane is $y = 0$.

1.8 Equation of a Plane Passing through three Points

2) The equation of the plane parallel of the XOY plane is $z =$ constant. Similarly, the equation of the plane parallel to the ZOX plane is $y =$ constant.

3) The equation of planes parallel to the x axis, y axis, and z axis is $by + cz + d = 0$, $ax + cz + d = 0$ and $ax + by + d = 0$ respectively.

8) Find the equation of the plane through the three points $(1, 1, 1)$, $(1, -1, 1)$, $(-7, -3, -5)$.

Sol. Let the required equation of the plane passing through the point $(1, 1, 1)$ be
$$a(x - 1) + b(y - 1) + c(z - 1) = 0. \tag{1.23}$$

It passes through $(1, 1, 1)$

$$\therefore a(0) + b(-2) + c(0) = 0 \Rightarrow b = 0$$

It passes through $(-7, -3, -5)$

$$\therefore -8a - 4b - 6c = 0 \therefore a = \frac{-3c}{4}$$

$\therefore$ The Equation (1.23) becomes;

$$\frac{-3c}{4}(x - 1) + c(z - 1) = 0$$
$$\therefore -3x + 3 + 4z - 4 = 0$$
$$\therefore 3x - 4z + 1 = 0.$$

9) Show that the points $(0, -1, 0)$, $(2, 1, -1)$, $(1, 1, 1)$ and $(3, 3, 0)$ are coplanar.

Sol. First, we will find the equation of a plane through three points $(0, -1, 0)$, $(2, 1, -1)$ and $(1, 1, 1)$.
Let the equation of the plane passing through $(0, -1, 0)$ is
$$a(x - 0) + b(y + 1) + c(z - 0) = 0. \tag{1.24}$$

It passes through point $(2, 1, -1)$

$$\therefore 2a + 2b - c = 0, \tag{1.25}$$

and it passes through point $(1, 1, 1)$

$$\therefore a + 2b + c = 0. \tag{1.26}$$

∴ By cross multiplication, we get

$$\frac{a}{\begin{vmatrix} 4 & -8 \\ 2 & -7 \end{vmatrix}} = \frac{b}{\begin{vmatrix} -8 & -6 \\ -7 & -7 \end{vmatrix}} = \frac{c}{\begin{vmatrix} -6 & 4 \\ -7 & 2 \end{vmatrix}}$$

$$\therefore \frac{a}{-28+16} = \frac{b}{56-42} = \frac{c}{-12+28}$$

$$\therefore \frac{a}{-12} = \frac{b}{14} = \frac{c}{16}$$

Substituting value of a, b and c in (1.24), we get

$$4x - 3(y+1) + 2z = 0$$

$$\therefore 4x - 3y + 2z - 3 = 0. \tag{1.27}$$

Four given points are coplanar if the fourth point $(3, 3, 0)$ lies on the plane (1.27).
i.e., $4(3) - 3(3) + 2(0) - 3 = 0$.
∴ The given four points are coplanar.

10) Show that the four points $(6, -4, 4)$, $(0, 0, -4)$ intersects the join of $(-1, -2, -3)$, $(1, 2, -5)$.

Sol. Let $P(6, -4, 4)$, $Q(0, 0, -4)$, $R(-1, -2, -3)$, and $S(1, 2, -5)$ be the given four points.
To prove that, the line joining P, Q intersects the line joining R, S.
The line PQ intersects RS, if the points P, Q, R, S lie on the same.
i.e., we have to prove that the given four points are coplanar.
Let the required equation of the plane passing through the point $P(6, -4, 4)$ is

$$a(x-6) + b(y+4) + c(z-4) = 0. \tag{1.28}$$

It passes through the point $(0, 0, -4)$

$$\therefore -6a + 4b - 8c = 0. \tag{1.29}$$

It passes through the point $(-1, -2, -3)$

$$\therefore -7a + 2b - 7c = 0. \tag{1.30}$$

∴ By cross multiplication, we get

$$\frac{a}{\begin{vmatrix} 4 & -8 \\ 2 & -7 \end{vmatrix}} = \frac{b}{\begin{vmatrix} -8 & -6 \\ -7 & -7 \end{vmatrix}} = \frac{c}{\begin{vmatrix} -6 & 4 \\ -7 & 2 \end{vmatrix}}$$

$$\therefore \frac{a}{-28+16} = \frac{b}{56-42} = \frac{c}{-12+28}$$

$$\therefore \frac{a}{-12} = \frac{b}{14} = \frac{c}{16}$$

Substituting the values of a, b, and c in (1.28), we get

$$-12(x-6) + 14(y+4) + 16(z-4) = 0$$

$$\therefore -12x + 72 + 14y + 56 + 16z - 64 = 0$$

$$\therefore -12x + 14y + 16z + 64 = 0$$

$$\therefore 6x - 7y - 8z - 32 = 0. \quad (1.31)$$

Four given points are coplanar if the fourth point $(1, 2, -5)$ lies on the plane (1.31).

$$\therefore 6(1) - 7(2) - 8(-5) - 32 = 6 - 14 + 40 - 32 = 0.$$

∴ P, Q, R, S are coplanar.
∴ PQ intersects RS.

11) If from the point $P(a, b, c)$, perpendiculars PL, MP be drawn to YZ and ZX planes, find the equation of the plane OLM.

Sol. The coordinate of L, the foot of the perpendicular from $P(a, b, c)$ on YZ plane $(x = 0)$ are $(0, b, c)$. The coordinate of M, the foot of the perpendicular from $P(a, b, c)$ on XZ plane $(y = 0)$ are $(a, 0, c)$.
∴ We have $L(0, b, c)$, $M(a, 0, c)$, and $O(0, 0, 0)$.
The equation of the plane through $O(0, 0, 0)$ is

$$A(x-0) + B(y-0) + C(z-0) = 0$$

$$\therefore Ax + By + Cz = 0. \quad (1.32)$$

It passes through the point $L(0, b, c)$

$$\therefore bB + cC = 0. \quad (1.33)$$

It passes through the point $M(a, 0, c)$

$$\therefore aA + cC = 0. \tag{1.34}$$

$\therefore$ By cross multiplication, we get

$$\frac{A}{bc} = \frac{B}{ac} = \frac{C}{-ab}.$$

Substituting the values of A, B, and C in (1.32), we get

$$bcx + acy - abz = 0 \therefore \frac{x}{a} + \frac{y}{b} - \frac{z}{c} = 0$$

is the required equation of the plane OLM.

12) Show that $(-1, 4, -3)$ is the circumcenter of the triangle formed by the points $(3, 2, -5), (-3, 8, -5)$, and $(-3, 2, 1)$.

Sol. Let $P(-1, 4, -3)$ is the circumcenter of the triangle formed by the points $A(3, 2, -5), B(-3, 8, -5)$, and $C(-3, 2, 1)$ if $PA = PB = PC$ and $P, A, B,$ and C are coplanar.

$$\therefore PA = \sqrt{(3+1)^2 + (2-4)^2 + (-5+3)^2} = 2\sqrt{6}$$

Similarly, $PB = 2\sqrt{6}, PC = 2\sqrt{6}$

$$\therefore PA = PB = PC.$$

Let the equation of the plane passing through the point $A(3, 2, 5)$ is

$$a(x-3) + b(y-2) + c(z+5) = 0. \tag{1.35}$$

It passes through the point $B(-3, 8, -5)$

$$\therefore -6a + 6b = 0 \Rightarrow a = b. \tag{1.36}$$

It passes through the point $C(-3, 2, -1)$

$$\therefore -6a + 6c = 0 \Rightarrow a = c. \tag{1.37}$$

Substituting in (1.35), we get

$$a(x-3) + a(y-2) + a(z+5) = 0$$

$$\therefore x + y + z = 0. \tag{1.38}$$

The point $P(-1, 4, -3)$ lies on the plane (1.38), if it satisfies equation (1.38).

i.e., $-1 + 4 - 3 = 0$.

$\therefore$ The points P, A, B, C are coplanar.

$\therefore$ Both the conditions for P to be the circumcenter of $\triangle ABC$ are satisfied.

$\therefore P$ is the circumcenter of $\triangle ABC$.

1.9 Equation of any Plane Parallel to a Given Plane

Let the given equation of the plane be

$$ax + by + cz + d = 0. \tag{1.39}$$

Let the equation of plane parallel to (1.39) be

$$a_1 x + b_1 y + c_1 z + d = 0. \tag{1.40}$$

Since the planes are parallel, we get

$$\frac{a_1}{a} = \frac{b_1}{b} = \frac{c_1}{c} = t$$

$$\therefore a_1 = at, b_1 = bt, c_1 = ct.$$

$\therefore$ Equation (1.40) becomes,

$$tax + tby + tcz + d = 0$$

$$\therefore ax + by + cz + \frac{d}{t} = 0$$

i.e., $ax + by + cz + d_1 = 0$ where $d_1 = \frac{d}{t}$.

$\therefore$ The equation of any plane parallel to the plane $ax + by + cz + d = 0$ is

$$ax + by + cz + d_1 = 0.$$

13) Find the equation of the plane through the point (x_1, y_1, z_1) and parallel to the plane $ax + by + cz + d = 0$.

16 Plane

Sol. The given equation of the plane is

$$ax + by + cz + d = 0, \qquad (1.41)$$

and let the given point be $P(x_1, y_1, z_1)$.
The equation of any plane parallel to the plane (1.41) is

$$ax + by + cz + d_1 = 0.$$

It passes through point $P(x_1, y_1 z_1)$ then $ax_1 + by_1 + cz_1 + d_1 = 0$.
i.e., $ax_1 + by_1 + cz_1 = -d_1$.
$\therefore$ The equation of any plane parallel to the plane (1.41) and passing through point (x_1, y_1, z_1) is given by

$$ax + by + cz - (ax_1 + by_1 + cz_1) = 0$$

$$\therefore a(x - x_1) + b(y - y_1) + c(z - z_1) = 0.$$

14) Find the equation of the plane through the point $(2, 3, 4)$ and parallel to the plane $5x - 6y + 7z = 3$.

Sol. The equation of any plane parallel to the plane $5x - 6y + 7z = 3$ and passing through the point $(2, 3, 4)$ is given by

$$5(x - 2) - 6(y - 3) + 7(z - 4) = 0$$

$$\therefore 5x - 6y + 7z - 20 = 0$$

is the required equation of the plane.

1.10 Equation of Plane Passing through the Intersection of Two Given Planes

Let

$$P_1 : a_1 x + b_1 y + c_1 z + d_1 = 0, \qquad (1.42)$$

and

$$P_2 : a_2 x + b_2 y + c_2 z + d_2 = 0, \qquad (1.43)$$

be the two given planes.
The equation of the plane passing through the intersection of two planes is given by $P_1 + \lambda P_2 = 0$; where λ is any constant.
Remark: If two planes are parallel then $P_1 = 0$ or $P_2 = 0$.

1.11 Equation of the Plane Passing through the Intersection of Two Given Planes and Passing through a Given Point

Let the equation of two given planes be

$$a_1x + b_1y + c_1z + d_1 = 0, \tag{1.44}$$

and

$$a_2x + b_2y + c_2z + d_2 = 0, \tag{1.45}$$

and let $P(x_1, y_1, z_1)$ be any given point.

The Equation of any plane passing through the intersection of the planes given by (1.44) and (1.45) can be represented as

$$(a_1x + b_1y + c_1z + d_1) + k(a_2x + b_2y + c_2z + d_2) = 0, \tag{1.46}$$

where k is any arbitrary constant.

Equation (1.46) passes through point $P(x_1, y_1, z_1)$

$$\therefore (a_1x_1 + b_1y_1 + c_1z_1 + d_1) + k(a_2x_1 + b_2y_1 + c_2z_1 + d_2) = 0$$

which gives the value of k.

Substituting the value of k in Equation (1.46), we get the required equation of the plane.

15) Find the equation of the plane through the intersection of the planes $x + y + z = 6$ and $2x + 3y + 4z + 5 = 0$ and the point $(1, 1, 1)$.

Sol. Let the equation of the plane through the intersection of the planes is

$$(x + y + z - 6) + k(2x + 3y + 4z + 5) = 0. \tag{1.47}$$

It passes through the point $(1, 1, 1)$

$$\therefore 20x + 23y + 26z - 69 = 0$$

is the required equation of the plane.

16) Find the equation of the plane through the intersection of the planes $x - 3y + 2z + 3 = 0$ and $3x - y - 2z - 5 = 0$ and the origin.

Sol. Let the equation of the plane through the intersection of the planes is

$$(x - 3y + 2z + 3) + k(3x - y - 2z - 5) = 0. \tag{1.48}$$

It passes through the origin.

$$\therefore 3 - 5k = 0 \Rightarrow k = \frac{3}{5}$$

Substituting the value of k in Equation (1.47), we get

$$(x - 3y + 2z + 3) + \frac{3}{5}(3x - y - 2z - 5) = 0$$

$$\therefore 14x - 18y + 4z = 0$$

$$\therefore 7x - 9y + 2z = 0$$

is the required equation of the plane.

17) Find the equation of the plane through the points $(1, -2, 4)$ and $(3, -4, 5)$ and parallel to the y axis.

Sol. Let the equation the plane parallel to y axis is

$$ax + cz + d = 0. \qquad (1.49)$$

It passes through the points $(1, -2, 4)$ and $(3, -4, 5)$
$\therefore a + 4c + d = 0$ and $3a + 5c + d = 0$
By cross multiplication,

$$\frac{a}{\begin{vmatrix} 4 & 1 \\ 5 & 1 \end{vmatrix}} = \frac{c}{\begin{vmatrix} 1 & 1 \\ 1 & 3 \end{vmatrix}} = \frac{d}{\begin{vmatrix} 1 & 4 \\ 3 & 5 \end{vmatrix}}$$

$$\therefore \frac{a}{-1} = \frac{c}{2} = \frac{d}{-7}$$

$$\therefore a = \frac{d}{7}, c = \frac{-2}{7}d$$

Substituting in Equation (1.49), we get $\frac{d}{7}x - \frac{8}{7}dz + d = 0$

$$\therefore x - 8z + 7 = 0$$

is the required equation of the plane.

18) Find the equation of the plane passing through $(-2, 3, 10)$ and through the z axis.

Sol. Let the equation of the plane through the z axis is $ax + by = 0$.

1.11 Equation of the Plane Passing through the Intersection

It passes through the point $(-2, 3, 10)$; we get $-2a + 3b = 0 \Rightarrow a = \frac{3b}{2}$
Substituting in equation $ax + by = 0$, we get $\frac{3b}{2}x + by = 0$

$$3x + 2y = 0$$

is the required equation of the plane.

19) Find the equation of the plane through the intersection of the planes $x + y + z = 1$, $2x + 3y + 5z = 5$, and perpendicular to the plane $x - y + z = 0$.

Sol. The equation of the plane through the intersection of planes is given by

$$(x + y + z - 1) + k(2x + 3y + 5z - 5) = 0. \tag{1.50}$$

The direction ratios of normal to the plane (1.50) are
$(1 + 2k, 1 + 3k, 1 + 5k)$, and the direction cosines of normal to the plane $x - y + z = 0$ are $1, -1, 1$.
Since plane (1.50) is perpendicular to $x - y + z = 0$, we get

$$(1 + 2k)(1) + (1 + 3k)(-1) + (1 + 5k)(1) = 0$$

$$\therefore 1 + 2k - 1 - 3k + 1 + 5k = 0$$

$$\therefore 4k + 1 = 0$$

$$\therefore k = \frac{-1}{4}.$$

Substituting the value of k in Equation (1.50), we get

$$(x + y + z - 1) - \tfrac{1}{4}(2x + 3y + 5z - 5) = 0$$
$$\therefore 2x + y + 9z + 1 = 0.$$

20) Find the equation of the plane passing through points $(1, 1, 2)$ and $(2, 4, 3)$ and perpendicular to the plane $x - 3y + 7z + 5 = 0$.

Sol. The equation of a plane passes through the point $(1,1,2)$ is

$$a(x - 1) + b(y - 1) + c(z - 2) = 0 \tag{1.51}$$

Since plane (1.51) passes through the point $(2, 4, 3)$, we get

$$a + 3b + c = 0. \tag{1.52}$$

Moreover, plane (1.51) is perpendicular to $x - 3y + 7z + 5 = 0$, we get

$$a - 3b - 7c = 0. \tag{1.53}$$

By cross multiplication, we get

$$\frac{a}{\begin{vmatrix} 3 & 1 \\ -3 & 7 \end{vmatrix}} = \frac{b}{\begin{vmatrix} 1 & 1 \\ 7 & 1 \end{vmatrix}} = \frac{c}{\begin{vmatrix} 1 & 3 \\ 1 & -3 \end{vmatrix}}$$

$$\therefore \frac{a}{24} = \frac{b}{-6} = \frac{c}{-6}$$
$$\therefore \frac{a}{-4} = \frac{b}{1} = \frac{c}{1} = t \Rightarrow a = -4t, b = t, c = t.$$

Substituting the value of a, b, c in Equation (1.51), we get $4x - y - z = 1$ which is the required equation of the plane.

21) Find the equation to the plane through the point $(-1, 3, 2)$ and perpendicular to the planes $x + 2y + 2z = 5$ and $3x + 3y + 2z = 8$.

Sol. The equation of the plane passes through the point $(-1, 3, 2)$

$$\therefore a(x+1) + b(y-3) + c(z-2) = 0. \tag{1.54}$$

Since plane (1.54) is perpendicular to the planes $x + 2y + 2z = 5$ and $3x + 3y + 2z = 8$, we get

$$a(1) + b(2) + c(2) = 0, \tag{1.55}$$

and
$$a(3) + b(3) + c(2) = 0. \tag{1.56}$$

By cross multiplication, we get $\frac{a}{-2} = \frac{b}{4} = \frac{c}{-3}$
Substituting the values of a, b, c in Equation (1.54), we get

$$2(x+1) - 4(y-3) + 3(z-2) = 0$$

$$\therefore 2x - 4y + 3z + 8 = 0$$

which is the required equation of the plane.

22) The plane $x - y + z = 2$ is rotated through a right angle about its line of intersection with the plane $2x + y - 3z = 1$. Find the equation of the plane in its new position.

Sol. The equation of the plane through the line of intersection of the plane $x - y + z = 2$ and $2x + y - 3z = 1$
is given by
$$(x - y + z - 2) + k(2x + y - 3z - 1) = 0$$

$$\therefore (1+2k)x + (-1+k)y + (1-3k)z - (2+k) = 0. \quad (1.57)$$

Since (1.57) is the right angle to the plane $x - y + z = 2$.

$$\therefore (1+2k)(1) + (-1+k)(-1) + (1-3k)(1) = 0$$

$$\therefore k = \frac{3}{2}$$

Substituting the value of k in Equation (1.57), we get

$$(1+3)x + \left(-1+\frac{3}{2}\right)y + \left(1-3\left(\frac{3}{2}\right)\right)z - \left(2+\left(\frac{3}{2}\right)\right) = 0$$

$$\therefore 4x + \frac{y}{2} - \frac{7}{2}z - \frac{7}{2} = 0$$

$$8x + y - 7z - 7 = 0$$

is the required equation of the plane.

1.12 Angle between Two Planes

Let two planes be given by

$$a_1 x + b_1 y + c_1 z + d_1 = 0, \quad (1.58)$$

$$a_2 x + b_2 y + c_2 z + d_2 = 0. \quad (1.59)$$

The angle between two planes is equal to the angle between their normal

The direction ratios of the normal to the plane (1.58) and (1.59) are $a_1, b_1, c_1,$ and a_2, b_2, c_2 respectively.

$\therefore$ The angle θ between the plane is given by

$$\cos \theta = \frac{a_1 a_2 + b_1 b_2 + c_1 c_2}{\sqrt{a_1^2 + b_1^2 + c_1^2}\sqrt{a_2^2 + b_2^2 + c_2^2}}.$$

Remark:
1) Condition of Perpendicularity:
If two planes are perpendicular then their normals are also perpendicular.
i.e., the normals are perpendicular if $a_1 a_2 + b_1 b_2 + c_1 c_2 = 0$.
$\therefore$ The planes are perpendicular if $a_1 a_2 + b_1 b_2 + c_1 c_2 = 0$.

2) Condition of Parallelism:
If two planes are parallel then their normals are also parallel.

i.e., $\dfrac{a_1}{a_2} = \dfrac{b_1}{b_2} = \dfrac{c_1}{c_2}.$

23) Find the angle between the planes $2x - y + 2z = 3$ and
$$3x + 6y + 2z = 4.$$

Sol. The given planes are $2x - y + 2z = 3$ and $3x + 6y + 2z = 4$.
The direction ratios of the normals to the planes are $2, -1, 2$, and $3, 6, 2$.
$\therefore$ The angle between the planes is given by

$$\cos\theta = \dfrac{2(3) + (-1)(6) + 2(2)}{\sqrt{4+1+4}\sqrt{9+36+4}} = \dfrac{4}{21}$$

$$\therefore \theta = \cos^{-1}\left(\dfrac{4}{21}\right).$$

24) Find the angle between the planes $3x - 4y + 5z = 0$ and $2x - y - 2z = 5$.

Sol. The given planes are $3x - 4y + 5z = 0$ and $2x - y - 2z = 5$.
$\therefore$ The direction ratios of the normals to the planes are $3, -4, 5$ and $2, -1, -2$.
$\therefore$ The angle between the planes is given by

$$\cos\theta = \dfrac{3(2) + (-4)(-1) + 5(-2)}{\sqrt{9+16+25}\sqrt{4+1+4}} = 0$$
$$\therefore \theta = \pi/2.$$

25) A plane passes through the point $(4, -1, 2)$ and is perpendicular to the line joining $(1, -5, 10)$, and $(2, 3, 4)$. Find the equation of the plane and the angles which it makes with the coordinate planes.

Sol. Let the equation of the plane passing through the point $P(4, -1, 2)$ be

$$a(x-4) + b(y+1) + c(z-2) = 0. \qquad (1.60)$$

The direction ratios of the line joining $(1, -5, 10)$ and $(2, 3, 4)$ are $-1, 8, -6$.

$$\therefore \dfrac{a}{+1} = \dfrac{b}{8} = \dfrac{c}{-6}$$

∴ Equation (1.60) becomes, $+1(x-4) + 8(y+1) - 6(z-2) = 0$

$$\therefore x + 8y - 6z + 16 = 0. \quad (1.61)$$

Direction ratios of normal to this plane are $1, 8, -6$.

Let the plane (1.61) make angle α, β, and γ with coordinate axes $x = 0$, $y = 0$ and $z = 0$ respectively.

$$\therefore \cos\alpha = \frac{(1)(1) + 8(0) + (-6)(0)}{\sqrt{(1)^2 + (8)^2 + (-6)^2}} = \frac{1}{\sqrt{101}} \Rightarrow \alpha = \cos^{-1}\left(\frac{1}{\sqrt{101}}\right)$$

Similarly,

$$\cos\beta = \frac{8}{\sqrt{101}} \Rightarrow \beta = \cos^{-1}\left(\frac{1}{\sqrt{101}}\right)$$

and

$$\cos\gamma = \frac{-6}{\sqrt{101}} \Rightarrow \gamma = \cos^{-1}\left(\frac{-6}{\sqrt{101}}\right).$$

1.13 Position of the Origin w.r.t. the Angle between Two Planes

Let the given planes be

$$a_1 x + b_1 y + c_1 z + d_1 = 0, \quad (1.62)$$

and

$$a_2 x + b_2 y + c_2 z + d_2 = 0, \quad (1.63)$$

where d_1, d_2 are both positive.

1. If the origin lies in the acute angle between the planes, then θ is obtuse i.e., $\cos\theta$ is negative or $a_1 a_2 + b_1 b_2 + c_1 c_2$ is negative.

2. If the origin lies in the obtuse angle between the planes the angle θ is acute i.e., $\cos\theta$ is positive or $a_1 a_2 + b_1 b_2 + c_1 c_2$ is positive.

26) Is the origin in the acute or obtuse angle between the planes $x + y - z = 3$ and $x - 2y + z = 3$.

Sol. Rewriting the given equation of the planes for making constant term positive, we get

$-x - y + z + 3 = 0$ and $-x + 2y - z + 3 = 0$.

$\therefore a_1 a_2 + b_1 b_2 + c_1 c_2 = (-1)(-1) + (-1)(2) + (1)(-1) = -2 < 0$.

$\therefore$ The origin lies in the acute angle between the planes.

1.14 Two Sides of a Plane

To determine a criterion for two given points to lie on the same or different sides of a given plane.

Let $P(x_1, y_1, z_1)$ and $Q(x_2, y_2, z_2)$ be two points that lie on the same or different sides of the plane

$$ax + by + cz + d = 0, \tag{1.64}$$

according to the expressions

$$ax_1 + by_1 + cz_1 + d \text{ and } ax_2 + by_2 + cz_2 + d,$$

are of the same sign or different signs.

Let the line PQ meet the given plane at R in the ratio $m = 1$ then the coordinates of R are

$$\left(\frac{mx_2 + x_1}{m+1}, \frac{my_2 + y_1}{m+1}, \frac{mz_2 + z_1}{m+1} \right)$$

lies on the same plane, we get

$$a\left(\frac{mx_2 + x_1}{m+1} \right) + b\left(\frac{my_2 + y_1}{m+1} \right) + c\left(\frac{mz_2 + z_1}{m+1} \right) + d = 0$$

$$\therefore m(ax_2 + by_2 + cz_2 + d) + (ax_1 + by_1 + cz_1 + d) = 0$$

$$\therefore m = -\left(\frac{ax_1 + by_1 + cz_1 + d}{ax_2 + by_2 + cz_2 + d} \right).$$

Case (i): If $ax_1 + by_1 + cz_1 + d$ and $ax_2 + by_2 + cz_2 + d$ are of opposite signs then m is positive.

$\therefore$ The plane (1.64) divides PQ internally in the ratio $m : 1$.

$\therefore P$ and Q lie on the opposite sides of the plane.

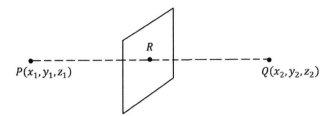

Case (ii): If $ax_1 + by_1 + cz_1 + d$ and $ax_2 + by_2 + cz_2 + d$ are of the same sign, then m is negative.

$\therefore PQ$ divides the plane (1.64) externally at R in the ratio $m = 1$.
$\therefore P$ and Q lies on the same side of the plane.

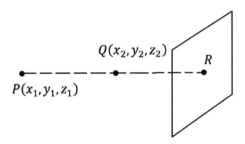

27) Show that the origin and the point $(2, -4, 3)$ lie on different sides of the plane $x + 3y - 5z + 7 = 0$.

Sol. The equation of the given plane is

$$x + 3y - 5z + 7 = 0. \qquad (1.65)$$

For the origin $(0, 0, 0)$;
    L.H.S of (1.65) becomes $(0) + 3(0) - 5(0) + 7 > 0$.
    For the point $(2, -4, 3)$;
    L.H.S of (1.65) becomes $2 + 3(-4) - 5(3) + 7 < 0$.
    Since the two signs are opposite; the given points lie on the opposite side of the given plane.

28) Prove that the points $(1, 2, 3)$ and $(0, 5, 1)$ on the same slide of the plane $y + z - 4 = 0$.

Sol. The equation of the given plane is

$$y + z - 4 = 0. \tag{1.66}$$

For the point $(1, 2, 3)$; L.H.S of (1.66) becomes $2 + 3 - 4 > 0$.

For the point $(0, 5, 1)$; L.H.S of (1.66) becomes $5 + 1 - 4 > 0$.

Since the two signs are the same; the given points lie on the same side of the given plane.

1.15 Length of the Perpendicular from a Point to a Plane

(a) To find the perpendicular distance of the point (x_1, y_1, z_1) from the plane $lx + my + nz = p$.

Let $P(x_1, y_1, z_1)$ be the given point and let $lx + my + nz = p$ be the plane ABC.

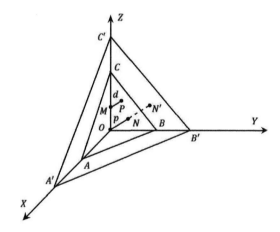

From P, draw PM perpendicular on the plane ABC and let $PM = d$ the required perpendicular distance.

From O draw ON perpendicular to the plane ABC so that $ON = P$ and direction cosines on ON are l, m, n.

Through P draw a plane $A^1B^1C^1$ parallel to the plane $A^1B^1C^1$ parallel to plane ABC to meet ON produced in N^1.

Then $ON^1 = ON + NN^1 = ON + MP = p + d$.

Also, the direction cosines of ON^1 are the same as those of ON

i.e., l, m, n

1.15 Length of the Perpendicular from a Point to a Plane

∴ Equation of the plane $A^1B^1C^1$ is $lx + my + nz = p + d$.
It passes through $P(x_1, y_1, z_1)$

$$\therefore lx_1 + my_1 + nz_1 = p + d$$

$$\therefore d = lx_1 + my_1 + nz_1 - p. \tag{1.67}$$

P and O are on the same side of the plane, then

$$d = -(lx_1 + my_1 + nz_1 - p). \tag{1.68}$$

Combining (1.67) and (1.68), the perpendicular distance formula is

$$d = \pm(lx_1 + my_1 + nz_1 - p).$$

(b) To find the perpendicular distance of the point (x_1, y_1, z_1) from the plane $ax + by + cz + d = 0$.
The equation of the given plane is

$$ax + by + cz + d = 0. \tag{1.69}$$

Dividing (1.69) by $\sqrt{a^2 + b^2 + c^2}$, equation (1.69) reduces to normal form,

$$\frac{a}{\sqrt{a^2 + b^2 + c^2}}x + \frac{b}{\sqrt{a^2 + b^2 + c^2}}y + \frac{c}{\sqrt{a^2 + b^2 + c^2}}z$$

$$+ \frac{d}{\sqrt{a^2 + b^2 + c^2}} = 0$$

$$\therefore \frac{a}{\sqrt{a^2 + b^2 + c^2}}x + \frac{b}{\sqrt{a^2 + b^2 + c^2}}y + \frac{c}{\sqrt{a^2 + b^2 + c^2}}z = \frac{-d}{\sqrt{a^2 + b^2 + c^2}}$$

which is the equation of the plane in the normal form

$$\frac{ax + by + cz + d}{\sqrt{a^2 + b^2 + c^2}} = 0$$

∴ Perpendicular distance of (x_1, y_1, z_1) from the plane is

$$p = \pm \frac{ax_1 + by_1 + cz_1 + d}{\sqrt{a^2 + b^2 + c^2}}.$$

Remark:

Distance of a point from a plane means perpendicular distance (if any other condition is not specified).

29) Find the distance of the point $(1, 1, 4)$ from the plane $3x - 6y + 2z + 11 = 0$.

Sol. The perpendicular distance of the point $(1, 1, 4)$ from the plane $3x - 6y + 2z + 11 = 0$ is given by

$$p = \pm \frac{ax_1 + by_1 + cz_1 + d}{\sqrt{a^2 + b^2 + c^2}}$$

$$\therefore p = \frac{3(1) - 6(1) + 2(4) + 11}{\sqrt{(3)^2 + (-6)^2 + (2)^2}} = \frac{16}{7}.$$

30) Show that distances between the parallel planes $2x - 2y + z + 3 = 0$ and $4x - 4y + 2z + 5 = 0$ is $\frac{1}{6}$.

Sol. The given planes

$$2x - 2y + z + 3 = 0 \qquad (1.70)$$

and $4x - 4y + 2z + 5 = 0$ are parallel.

∴ The distance between two parallel planes is

$$\frac{|p - p_1|}{\sqrt{a^2 + b^2 + c^2}} = \frac{|3 - 5/2|}{\sqrt{(2)^2 + (2)^2 + (1)^2}} = \frac{1/2}{3} = \frac{1}{6}.$$

1.16 Bisectors of Angles between Two Planes

To find the equation of the bisectors of the angle between the planes

$$ax + by + cz + d = 0$$

$$a_1 x + b_1 y + c_1 z + d_1 = 0.$$

Let $P(x, y, z)$ be a point on any one of the planes bisecting the angles between the planes.
∴ The perpendicular from $P(x, y, z)$ to the two planes must be equal.

i.e., $$\frac{ax + by + cz + d}{\sqrt{a^2 + b^2 + c^2}} = \frac{a_1 x + b_1 y + c_1 z + d_1}{\sqrt{a_1^2 + b_1^2 + c_1^2}}$$

are the equations of the two bisecting planes.

1.16 Bisectors of Angles between Two Planes

Remark:

1) Of these two bisecting planes, one bisects the acute and the other the obtuse angle between the planes.

2) The bisector of the acute angle makes with either of the planes an angle which is less than 45 and the bisector of the obtuse angle makes with either of them an angle which is greater than 45.

3) Position of origin w.r.t two non-parallel planes: Let θ be the angle between the two normal drawn from the origin to the two given planes then

$$\cos\theta = \frac{a_1a_2 + b_1b_2 + c_1c_2}{\sqrt{a_1^2 + b_1^2 + c_1^2}\sqrt{a_2^2 + b_2^2 + c_2^2}}$$

(a) If $a_1a_2 + b_1b_2 + c_1c_2 < 0$ then θ is obtuse then the angle between the planes is $(\pi - \theta)$ must be acute.

i.e., Origin must lie in the acute angle made by the planes.

(b) If $a_1a_2 + b_1b_2 + c_1c_2 > 0$ then θ is acute then the angle between the planes is $(\pi - \theta)$ will be obtuse.

i.e., Origin must lie in the obtuse angle made by the planes.

If d_1 and d_2 are both positive, the origin will be in the acute or obtuse angle between the planes $a_1x + b_1y + c_1 + d_1 = 0$ and $a_2x + b_2y + c_2z + d_2 = 0$.

i.e., $a_1a_2 + b_1b_2 + c_1c_2 <$ or > 0.

31) Determine whether the origin lies inside the acute or obtuse angle formed by the planes $x - 2y + 3z - 5 = 0$ and $2x - y - z + 3 = 0$.

Sol. Rewriting the equations of planes to make constant terms positive, we get

$$-x + 2y - 3z + 5 = 0 \text{ and } 2x - y - z + 3 = 0$$

$\therefore a_1a_2 + b_1b_2 + c_1c_2 = (-1)(2) + (2)(-1) + (-3)(-1) = -1 < 0.$

$\therefore$ The origin lies in the acute angle between two given planes.

32) Show that the plane $14x - 8y + 13 = 0$ bisects the obtuse angle between the planes $3x + 4y - 5z + 1 = 0$ and $5x + 12y - 13z = 0$.

Sol. The equations of the two bisecting planes are

$$\frac{3x + 4y - 5z + 1}{\sqrt{9 + 16 + 25}} = \pm \frac{5x + 12y - 13z}{\sqrt{25 + 144 + 169}}$$

$$\therefore \frac{3x + 4y - 5z + 1}{5\sqrt{2}} = \pm \frac{5x + 12y - 13z}{13\sqrt{2}}$$

$$\therefore 13(3x + 4y - 5z + 1) = \pm 5(5x + 12y - 13z)$$

$$\therefore 39x + 52y - 65z + 13 = \pm(25x + 60y - 65z)$$

$\therefore 14x - 8y + 13 = 0$ or $64x + 112y - 130z + 13 = 0$.

Let ϕ be the angle between $3x + 4y - 5z + 1 = 0$ and the bisector $14x - 8y + 13 = 0$ then

$$\cos\phi = \frac{(14)(3) + 4(-8)}{\sqrt{(3)^2 + (4)^2 + (-5)^2}\sqrt{(14)^2 + (-8)^2}} = \frac{10}{10\sqrt{130}} = \frac{1}{\sqrt{130}}$$

$$\therefore \tan\phi = \sqrt{\sec^2\theta - 1} = \sqrt{130 - 1} = \sqrt{129} > 1$$

$$\therefore \Phi > \frac{\pi}{4}.$$

$\therefore$ The angle between the two planes $3x + 4y - 5z + 1 = 0$ and $14x - 8y + 13 = 0$ is greater than $45°$.

$\therefore$ The bisecting plane $14x - 8y + 13 = 0$ is the bisector of the obtuse angle between the given planes.

33) Show that the origin lies in the acute angle between the planes $x + 2y + 2z = 9$ and $4x - 3y + 12z + 13 = 0$. Find the planes bisecting the angles between them and point out which bisects the acute angle.

Sol. Rewriting the given equation of the plane

$$-x - 2y - 2z + 9 = 0 \text{ and } 4x - 3y + 12z + 13 = 0.$$

The equations of two bisecting planes are

$$\frac{-x - 2y - 2z + 9}{\sqrt{(-1)^2 + (-2)^2 + (-2)^2}} = \pm \frac{4x - 3y + 12z + 13}{\sqrt{(4)^2 + (-3)^2 + (12)^2}}$$

$$\therefore \frac{-x - 2y - 2z + 9}{3} = \pm \frac{4x - 3y + 12z + 13}{13}$$

$\therefore -13x - 26y + 26z + 117 = \pm(12x - 9y + 36z + 39)$

$\therefore 25x + 17y + 62z - 78 = 0$ or $x + 35y - 10z - 156 = 0$.

Let ϕ be the angle between $x + 2y + 2z - 9 = 0$ and the bisecting plane $25x + 17y + 62z - 78 = 0$ then

$$\cos\phi = \frac{25(1) + 17(2) + 62(2)}{\sqrt{1+4+4}\sqrt{(25)^2 + (17)^2 + (62)^2}} = \frac{61}{68}$$

$$\tan\phi = \sqrt{\sec^2\phi - 1} = \sqrt{\left(\frac{68}{61}\right)^2 - 1} = \frac{\sqrt{903}}{61} < 1$$

$$\therefore \Phi < \frac{\pi}{4}.$$

$\therefore$ The bisecting plane $25x + 17y + 62z - 78 = 0$ is the bisector of the acute angle.

$\therefore$ The origin lies in the acute angle.

1.17 Pair of Planes

To find an equation that will be satisfied if and only if a point lies on either of the two planes. i.e., either on one plane or the other or both.

Let the equations of two planes be

$$P_1 : ax + by + cz + d = 0,$$

and

$$P_2 : a_1 x + b_1 y + c_1 z + d_1 = 0$$

then any point on $P_1 = 0$ or $P_2 = 0$ will also satisfy $P_1 P_2 = 0$.

$\therefore P_1 P_2 = 0$ is the equation of a pair of planes representing two planes $P_1 = 0$ and $P_2 = 0$.

i.e., $(ax + by + cz + d)(a_1 x + b_1 y + c_1 z + d_1) = 0$

Let point (x_1, y_1, z_1) lie on P_1.

$$\Rightarrow ax_1 + by_1 + cz_1 + d = 0$$

$$\therefore (ax_1 + by_1 + cz_1 + d)(a_1 x + b_1 y + c_1 z + d_1) = 0$$

Let point (x_1, y_1, z_1) lie on P_2.

$$\Rightarrow a_1 x_1 + b_1 y_1 + c_1 z_1 + d_1 = 0$$

$$\therefore (ax_1 + by_1 + cz_1 + d)(a_1x_1 + b_1y_1 + c_1z + d_1) = 0$$
$\Rightarrow$ the Point lies on $P_1P_2 = 0$
i.e., $ax_1 + by_1 + cz_1 + d = 0$ or $a_1x_1 + b_1y_1 + c_1z_1 + d_1 = 0$
$\Rightarrow (x_1, y_1, z_1)$ lies on P_1 or P_2.
Thus, a point (x, y, z) either lies on the plane P_1 or the plane P_2.
$\therefore (ax + by + cz + d)(a_1x + b_1y + c_1z + d_1) = 0$ is the equation of two planes, which is the general equation of second degree in x, y, z.

To find the condition that a homogeneous equation of second degree in x, y, z may represent a pair of planes: Let homogeneous second-degree equation in x, y, z be given by

$$ax^2 + by^2 + cz^2 + 2fyz + 2gzx + 2hxy = 0.$$

Let it represents two planes

$$l_1x + m_1y + n_1z = 0$$

$$l_2x + m_2y + n_2z = 0$$

$$\therefore ax^2 + by^2 + cz^2 + 2fyz + 2gzx + 2hxy = (l_1x + m_1y + n_1z)(l_2x + m_2y + n_2z).$$

Now equating coefficients, we get

$$l_1l_2 = a, m_1m_2 = b, n_1n_2 = c, l_1m_2 + l_2m_1 = 2h, m_1n_2 + m_2n_1 = 2f$$

and $n_1l_2 + n_2l_1 = 2g.$

The required condition is essentially the condition for the consistency of these equations is obtained on eliminating $l_1, m_1, n_1; l_2, m_2, n_2$ from the above six relations and this can be easily affected as follows.

$$\begin{vmatrix} l_1 & l_2 & 0 \\ m_1 & m_2 & 0 \\ n_1 & n_2 & 0 \end{vmatrix} \times \begin{vmatrix} l_2 & l_1 & 0 \\ m_2 & m_1 & 0 \\ n_2 & n_1 & 0 \end{vmatrix} = 0$$

$$\begin{vmatrix} 2l_1l_2 & l_1m_2 + l_2m_1 & l_1n_2 + l_2n_1 \\ l_1m_2 + l_2m_1 & 2m_1m_2 & m_1n_2 + m_2n_1 \\ l_1n_2 + l_2n_1 & m_1n_2 + m_2n_1 & 2n_1n_2 \end{vmatrix} = 0$$

$$\Rightarrow \begin{vmatrix} 2a & 2h & 2g \\ 2h & 2b & 2f \\ 2g & 2f & 2c \end{vmatrix} = 0$$

$$\Rightarrow 8\begin{vmatrix} a & h & g \\ h & b & f \\ g & f & c \end{vmatrix} = 0$$

$$\Rightarrow abc + 2fgh - af^2 - bg^2 - ch^2 = 0$$

which is the required sufficient condition.
i.e., If the equation $ax^2 + by^2 + cz^2 + 2fyz + 2gzx + 2hxy = 0$ represents two planes, we have the condition

$$abc + 2fgh - af^2 - by^2 - ch^2 = 0.$$

Let $abc + 2fgh - af^2 - bg^2 - ch^2 = 0$ holds.
Now, the given equation can be expressed in terms of x as

$$x = \frac{-(gz + hy) \pm \sqrt{(gz + hy)^2 - a(by^2 + cz^2 + 2fyz)}}{a}; a \neq 0.$$

Now, if the given condition is satisfied, then the expression under the radical sign must be a perfect square. So, the given homogeneous equation can be expressed as a product of two linear equations, and hence it will represent a pair of planes.

Corollary: Angle between planes:

If θ be the angle between the planes represented by the equation

$$ax^2 + by^2 + cz^2 + 2fyz + 2gzx + 2hxy = 0$$

we have if $l_1 l_2 + m_1 m_2 + n_1 n_2 \neq 0$ then

$$\tan \theta = \pm \frac{\sqrt{(m_1 n_2 - m_1 n_1)^2 + (n_1 l_2 + n_2 l_1)^2 + (l_1 m_2 - l_2 m_1)^2}}{l_1 l_2 + m_1 m_2 + n_1 n_2}$$

$$= \pm \frac{2\sqrt{f^2 + g^2 + h^2 - ab - bc - ca}}{a + b + c}.$$

Remark:

1) If the two planes are orthogonal; i.e., right-angled to each other then $\theta = 90$ or $\cos \theta = 0$ or $a + b + c = 0$.

2) If $\theta = 0$ or 180, $\tan \theta = 0$ then $f^2 + g^2 + h^2 = ab + bc + ca$, and then the two planes are parallel.

3) If $\tan\theta \neq 0$; i.e., $f^2 + g^2 + h^2 \neq ab + bc + ca$, then the two planes are intersecting.

34) Show that equation $12x^2 - 2y^2 - 6z^2 - 2xy + 7yz + 6zx = 0$ represents a pair of planes and also finds the angle between each plane.

Sol. The given equation is $12x^2 - 2y^2 - 6z^2 - 2xy + 7yz + 6zx = 0$ comparing with $ax^2 + by^2 + cz^2 + 2fyz + 2gzx + 2hxy = 0$ we get,
$$a = 12, b = -2, c = -6, 2f = 7, 2g = 6, 2h = -2$$

Condition for the homogeneous second-degree equation to represent a pair of planes is
$$abc + 2fgh - af^2 - bg^2 - ch^2 = 0.$$

$\therefore 12x^2 - 2y^2 - 6z^2 - 2xy + 7yz + 6zx = 0$ represents a pair of planes.
Let θ be the angle between

$$\tan\theta = \frac{2\sqrt{(f^2+g^2+h^2-bc-ca-ab)}}{a+b+c}$$
$$= \frac{2\sqrt{(7/2)^2+(3)^2+(-1)^2-(-2)(-6)-(-6)(12)-(12)(-2)}}{12-2-6}$$
$$= \frac{2(21/2)}{4}$$
$$\therefore \theta = \tan^{-1}\left(\frac{21}{4}\right).$$

35) Show that the equation $\frac{a}{y-z} + \frac{b}{z-x} + \frac{c}{x-y} = 0$ represents a pair of planes.

Sol. Given equation is
$$\frac{a}{y-z} + \frac{b}{z-x} + \frac{c}{x-y} = 0.$$

Let $y - z = X$; $z - x = Y$ and $x - y = Z$
$\Rightarrow X + Y + Z = 0$ or $Z = -(X+Y)$
$\therefore$ Given equation becomes
$$\frac{a}{X} + \frac{b}{Y} + \frac{c}{Z} = 0$$
i.e., $\quad \dfrac{a}{X} + \dfrac{b}{Y} - \dfrac{c}{X+Y} = 0$
$\therefore ay(-X-Y) + bX(-X-Y) + cXY = 0$
$\therefore aXY - aY^2 - bX^2 - bXY + cXY = 0$
$\therefore bX^2 + aY^2 + (a+b-c)XY = 0$
which is a homogenous equation of second degree in XY.

∴ The given equation represents a pair of planes.

1.18 Orthogonal Projection on a Plane

The foot of the perpendicular from a point to a given plane is called the orthogonal projection of the point on the plane. The plane on which we project is called the plane of the projection.

Remarks:

1) The projection of a curve on a plane is the locus of the projections on the plane of any point on the curve.

2) The projection on a given plane of the area enclosed by a plane curve is the area enclosed by the projection of the curve on the plane.

3) The projection of a straight line on a given plane is the locus of the feet of the perpendiculars drawn from points on the line on the plane.

Remarks:

1) The projection of a straight line is a straight line.

2) If a segment AB is parallel to the plane of projection, then the length of the projection is the same as that of AB.

3) The projection of the area A, enclosed by a curve in a plane is $A\cos\theta$ where θ is the angle between the plane of the curve containing the given area and the plane of projection.

4) If Ax, Ay, Az be the areas of projections of on area A, on the three coordinate planes then $A^2 = Ax^2 + Ay^2 + Az^2$.

36) Find the areas of the triangles whose vertices are the points $(a, 0, 0), (0, b, 0), (0, 0, c)$.

Sol. The vertices of the projection of the triangle on the XY plane are $(a, 0, 0), (0, b, 0),$ and $(0, 0, c)$.

$$Ax = \frac{1}{2}\begin{vmatrix} a & 0 & 1 \\ 0 & b & 1 \\ 0 & 0 & 1 \end{vmatrix} = \frac{1}{2}(ab)$$

Similarly,

$$Ay = \frac{1}{2}\begin{vmatrix} a & 0 & 1 \\ 0 & 0 & 1 \\ 0 & c & 1 \end{vmatrix} = \frac{1}{2}(-ac)$$

$$Az = \frac{1}{2}\begin{vmatrix} 0 & 0 & 1 \\ b & 0 & 1 \\ 0 & c & 1 \end{vmatrix} = \frac{1}{2}(bc).$$

$\therefore$ The area of the triangle is given by

$$A = \sqrt{Ax^2 + Ay^2 + Az^2} = \frac{1}{2}\sqrt{a^2b^2 + a^2c^2 + b^2c^2}.$$

1.19 Volume of a Tetrahedron

To find the volume of a tetrahedron when four vertices (x_1, y_1, z_1), (x_2, y_2, z_2), (x_3, y_3, z_3) and (x_4, y_4, z_4) are known.

Let $ABCD$ be a tetrahedron joining the points $A(x_1, y_1, z_1)$, $B(x_2, y_2, z_2)$, $C(x_3, y_3, z_3)$ and $D(x_4, y_4, z_4)$ then the volume of the tetrahedron is given by

$$V = \tfrac{1}{3}.AN \triangle BCD$$
$$= \tfrac{1}{3}p.\Delta$$

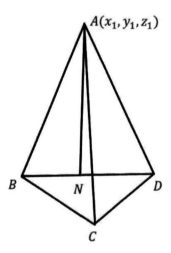

where Δ is the area of the triangle BCD.

1.19 Volume of a Tetrahedron

Let $\Delta x, \Delta y, \Delta z$ be the projections of $\triangle BCD$ on co-ordinate planes $x = 0, y = 0, z = 0$ respectively then

$$\Delta_x = \frac{1}{2}\begin{vmatrix} y_2 & z_2 & 1 \\ y_3 & z_3 & 1 \\ y_4 & z_4 & 1 \end{vmatrix}; \Delta_y = \frac{1}{2}\begin{vmatrix} x_2 & z_2 & 1 \\ x_3 & z_3 & 1 \\ x_4 & z_4 & 1 \end{vmatrix}; \Delta_z = \frac{1}{2}\begin{vmatrix} x_2 & y_2 & 1 \\ x_3 & y_3 & 1 \\ x_4 & y_4 & 1 \end{vmatrix}$$

$$\Delta = \Delta_x^2 + \Delta_y^2, \Delta_z^2$$

The equation of the plane BCD is given by

$$\begin{vmatrix} x & y & z & 1 \\ x_2 & y_2 & z_2 & 1 \\ x_3 & y_3 & z_3 & 1 \\ x_4 & y_4 & z_4 & 1 \end{vmatrix} = 0$$

$$\therefore x\begin{vmatrix} y_2 & z_2 & 1 \\ y_3 & z_3 & 1 \\ y_4 & z_4 & 1 \end{vmatrix} - y\begin{vmatrix} x_2 & z_2 & 1 \\ x_3 & z_3 & 1 \\ x_4 & z_4 & 1 \end{vmatrix} + z\begin{vmatrix} x_2 & y_2 & 1 \\ x_3 & y_3 & 1 \\ x_4 & y_4 & 1 \end{vmatrix} - \begin{vmatrix} x_2 & y_2 & z_2 \\ x_3 & y_3 & z_3 \\ x_4 & y_4 & z_4 \end{vmatrix} = 0$$

$\therefore$ The length of the perpendicular $p = AN$

$$\therefore p = x_1 \frac{x_1\begin{vmatrix} y_2 & z_2 & 1 \\ y_3 & z_3 & 1 \\ y_4 & z_4 & 1 \end{vmatrix} - y_1\begin{vmatrix} x_2 & z_2 & 1 \\ x_3 & z_3 & 1 \\ x_4 & z_4 & 1 \end{vmatrix} + z_1\begin{vmatrix} x_2 & y_2 & 1 \\ x_3 & y_3 & 1 \\ x_4 & y_4 & 1 \end{vmatrix} - \begin{vmatrix} x_2 & y_2 & z_2 \\ x_3 & y_3 & z_3 \\ x_4 & y_4 & z_4 \end{vmatrix}}{\sqrt{4(\Delta_x^2 + \Delta_y^2 + \Delta_z^2)}}$$

$$= \frac{\begin{vmatrix} x_1 & y_1 & z_1 & 1 \\ x_2 & y_2 & z_2 & 1 \\ x_3 & y_3 & z_3 & 1 \\ x_4 & y_4 & z_4 & 1 \end{vmatrix}}{2\Delta} = \frac{D}{2\Delta}$$

where $D = \begin{vmatrix} x_1 & y_1 & z_1 & 1 \\ x_2 & y_2 & z_2 & 1 \\ x_3 & y_3 & z_3 & 1 \\ x_4 & y_4 & z_4 & 1 \end{vmatrix}$

$$\therefore V = \tfrac{1}{3}\Delta = \tfrac{1}{3}\cdot\tfrac{D}{2\Delta}\cdot\Delta = \tfrac{1}{6}D = \tfrac{1}{6}\begin{vmatrix} x_1 & y_1 & z_1 & 1 \\ x_1 & y_2 & z_2 & 1 \\ x_3 & y_3 & z_3 & 1 \\ x_4 & y_4 & z_4 & 1 \end{vmatrix}$$

$$\therefore V = \frac{-D^3}{6D_1 D_2 D_3 D_4}$$

$$\therefore V = \left| \frac{D^3}{6D_1 D_2 D_3 D_4} \right|.$$

To find the volume of the tetrahedron when the equations of four faces are given.

Let the equations of form faces be given by

$$a_1 x + b_1 y + c_1 z + d_1 = 0, \qquad (1.71)$$

$$a_2 x + b_2 y + c_2 z + d_2 = 0, \qquad (1.72)$$

$$a_3 x + b_3 y + c_3 z + d_3 = 0, \qquad (1.73)$$

$$a_4 x + b_4 y + c_4 z + d_4 = 0. \qquad (1.74)$$

$$\text{Let } D = \begin{vmatrix} a_1 & b_1 & c_1 & d_1 \\ a_2 & b_2 & c_2 & d_2 \\ a_3 & b_3 & c_3 & d_3 \\ a_4 & b_4 & c_4 & d_4 \end{vmatrix}$$

Solving three Equations of (1.71) to (1.74) in turns, we get the four vertices

Let us solve the Equations (1.72), (1.73), and (1.74) to get one vertex.
$\therefore$ By Cramer's law,

$$\frac{x}{D_1^{(1)}} = \frac{y}{D_1^{(2)}} = \frac{z}{D_1^{(3)}} = \frac{-1}{D_1}$$

where $D_1 = - \begin{vmatrix} a_2 & b_2 & c_2 \\ a_3 & b_3 & c_3 \\ a_4 & b_4 & c_4 \end{vmatrix}$ = cofactor of d_1 in D

$$D_1^{(1)} = \begin{vmatrix} d_2 & b_2 & c_2 \\ d_3 & b_3 & c_3 \\ d_4 & b_4 & c_4 \end{vmatrix} = \begin{vmatrix} b_2 & c_2 & d_2 \\ b_3 & c_3 & d_3 \\ b_4 & c_4 & d_4 \end{vmatrix} = A_1$$

$$D_1^{(2)} = \begin{vmatrix} a_2 & d_2 & c_2 \\ a_3 & d_3 & c_3 \\ a_4 & d_4 & c_4 \end{vmatrix} = - \begin{vmatrix} a_2 & c_2 & d_2 \\ a_3 & c_3 & d_3 \\ a_4 & c_4 & d_4 \end{vmatrix} = B_1$$

1.19 Volume of a Tetrahedron

$$D_1^{(3)} = \begin{vmatrix} a_2 & b_2 & d_2 \\ a_3 & b_3 & d_3 \\ a_4 & b_4 & d_4 \end{vmatrix} = C_1$$

$$\therefore x = \frac{-A_1}{D_1}; y = \frac{-B_1}{D_1}; z = \frac{-C_1}{D_1}$$

where A_i, B_i, C_i, D_i are cofactors of a_i, b_i, c_i, d_i respectively in D.

$\therefore$ Coordinates of the corresponding vertex will be $\left(\frac{-A_1}{D_1}, \frac{-B_1}{D_1}, \frac{-C_1}{D_1}\right)$.

Similarly, the coordinates of the other vertices will be $\left(\frac{-A_2}{D_2}, \frac{-B_2}{D_2}, \frac{-C_2}{D_2}\right)$;

$\left(\frac{-A_3}{D_3}, \frac{-B_3}{D_3}, \frac{-C_3}{D_3}\right)$ and $\left(\frac{-A_4}{D_4}, \frac{-B_4}{D_4}, \frac{-C_4}{D_4}\right)$

$\therefore$ The volume of the tetrahedron will be

$$V = \frac{1}{6} \begin{vmatrix} \frac{-A_1}{D_1} & \frac{-B_1}{D_1} & \frac{-C_1}{D_1} & 1 \\ \frac{-A_2}{D_2} & \frac{-B_2}{D_2} & \frac{-C_2}{D_2} & 1 \\ \frac{-A_3}{D_3} & \frac{-B_3}{D_3} & \frac{-C_3}{D_3} & 1 \\ \frac{-A_4}{D_4} & \frac{-B_4}{D_4} & \frac{-C_4}{D_4} & 1 \end{vmatrix} = \frac{-1}{6 D_1 D_2 D_3 D_4} \begin{vmatrix} A_1 & B_1 & C_1 & D_1 \\ A_2 & B_2 & C_2 & D_2 \\ A_3 & B_3 & C_3 & D_3 \\ A_4 & B_4 & C_4 & D_4 \end{vmatrix}.$$

37) The vertices of a tetrahedron are $(0, 1, 2), (3, 0, 1) (4, 3, 6), (2, 3, 2)$; show that its volume is 6.

Sol. The volume of the tetrahedron is given by

$$V = \frac{1}{6} \begin{vmatrix} x_1 & y_1 & z_1 & 1 \\ x_2 & y_2 & z_2 & 1 \\ x_3 & y_3 & z_3 & 1 \\ x_4 & y_4 & z_4 & 1 \end{vmatrix} = \frac{1}{6} \begin{vmatrix} 0 & 1 & 2 & 1 \\ 3 & 0 & 1 & 1 \\ 4 & 3 & 6 & 1 \\ 2 & 3 & 2 & 1 \end{vmatrix}$$

$$= \frac{1}{6} \left[-1 \begin{vmatrix} 3 & 1 & 1 \\ 4 & 6 & 1 \\ 2 & 2 & 1 \end{vmatrix} + 2 \begin{vmatrix} 3 & 0 & 1 \\ 4 & 3 & 1 \\ 2 & 3 & 1 \end{vmatrix} - 1 \begin{vmatrix} 3 & 0 & 1 \\ 4 & 3 & 6 \\ 2 & 3 & 2 \end{vmatrix} \right]$$

$$= \frac{1}{6}[-1(6) + 2(6) - 1(-30)] = \frac{1}{6}(-6 + 12 + 30) = \frac{1}{6}(36) = 6.$$

38) A variable plane makes with the coordinate planes a tetrahedron of constant volume k^3. Show that the locus of the foot of the perpendicular from the origin to the plane is $(x^2 + y^2 + z^2)^3 = 6k^3 xyz$.

40 Plane

Sol. Let the equation of the plane be

$$\frac{x}{a}+\frac{y}{b}+\frac{z}{c}=1 \qquad (1.75)$$

∴ The four vertices of the tetrahedron are given by $(0,0,0),(0,0,0)$, $(0,b,0),(0,0,c)$.
∴ The volume of the tetrahedron is

$$V = \frac{1}{6}\begin{vmatrix} a & 0 & 0 & 1 \\ 0 & b & 0 & 1 \\ 0 & 0 & c & 1 \\ 0 & 0 & 0 & 1 \end{vmatrix} = \frac{abc}{6}$$

Given,

$$\frac{abc}{6}=k^3 \Rightarrow abc=6k^3. \qquad (1.76)$$

Let (α,β,γ) be the coordinates of the foot of the perpendicular from the origin then

$$\frac{\alpha}{a}+\frac{\beta}{b}+\frac{\gamma}{c}=1 \qquad (1.77)$$

and $\frac{\alpha-0}{\frac{1}{a}}=\frac{\beta-0}{\frac{1}{b}}=\frac{\gamma-0}{\frac{1}{c}}=\lambda$
i.e., $a\alpha = b\beta = c\gamma = \lambda$

$$\therefore a=\frac{\lambda}{\alpha};\ b=\frac{\lambda}{\beta}\text{ and } c=\frac{\lambda}{\gamma} \qquad (1.78)$$

By Equation (1.77);

$$\frac{1}{\lambda}\left(\alpha^2+\beta^2+\gamma^2\right)=1$$

i.e., $\alpha^2+\beta^2+\gamma^2=\lambda \qquad (1.79)$

By Equations (1.78) and (1.79), we get
$a=\frac{\left(\alpha^2+\beta^2+\gamma^2\right)}{\alpha};\ b=\frac{\left(\alpha^2+\beta^2+\gamma^2\right)}{\beta}$ and $c=\frac{\left(\alpha^2+\beta^2+\gamma^2\right)}{\gamma}$
∴ Equation (1.76) becomes;

$$\frac{\left(\alpha^2+\beta^2+\gamma^2\right)^3}{\alpha\beta\gamma}=6k^3$$

$$\therefore \left(\alpha^2+\beta^2+\gamma^2\right)^3 = 6k^3\alpha\beta\gamma$$

1.19 Volume of a Tetrahedron

∴ The locus of the foot of the perpendicular will be

$$(x^2 + y^2 + z^2)^3 = 6k^3 xyz.$$

39) A variable plane makes with the coordinate planes a tetrahedron of constant volume $64\lambda^3$. Find

1) The locus of the centroid of the tetrahedron and

2) The locus of the foot of the perpendicular from the origin to the plane.

Sol. Let the equation of the variable plane be

$$ax + by + cz = d. \qquad (1.80)$$

The points A, B, and C where the plane (1.80) meets the coordinate axes are given by $\left(\frac{d}{a}, 0, 0\right)$, $\left(0, \frac{d}{b}, 0\right)$ and $\left(0, 0, \frac{d}{c}\right)$.
∴ The volume of the tetrahedron $OABC$ will be

$$V = \frac{1}{6} \begin{vmatrix} \frac{d}{a} & 0 & 0 \\ 0 & \frac{d}{b} & 0 \\ 0 & 0 & \frac{d}{c} \end{vmatrix} = \frac{d^3}{6abc}$$

Given

$$\frac{d^3}{6abc} = 64\lambda^3 \qquad (1.81)$$

1) Let (α, β, γ) be the coordinates of the centroid of the tetrahedron then

$$\alpha = \frac{0 + \frac{d}{a} + 0 + 0}{4} \Rightarrow a = \frac{d}{4\alpha}$$

Similarly, $b = \frac{d}{4\beta}$ and $c = \frac{d}{4\gamma}$.

Substituting the values of a, b, c in equation (1.81), we get $\alpha\beta\gamma = 6\lambda^3$
∴ The locus of (α, β, γ) is given by $xyz = 6\lambda^3$.

2) Let (x_1, y_1, z_1) be the coordinates of the foot of the perpendicular from $(0, 0, 0)$ to the plane, then its direction cosines are proportional to (x_1, y_1, z_1). Since the direction cosines to the normal to the plane (1.80) are proportional to a, b, c, we get

$$\frac{x_1}{a} = \frac{y_1}{b} = \frac{z_1}{c}. \qquad (1.82)$$

Since (x_1, y_1, z_1) lies on the plane (1.80), we get

$$ax_1 + by_1 + cz_1 = d. \tag{1.83}$$

From Equation (1.82), we get

$$\frac{x_1^2}{ax_1} = \frac{y_1^2}{by_1} = \frac{z_1^2}{cz_1} = \frac{x_1^2+y_1^2+z_1^2}{ax_1+by_1+cz_1} = \frac{x_1^2+y_1^2+z_1^2}{d}$$

$$\therefore a = \frac{dx_1}{x_1^2+y_1^2+z_1^2}; b = \frac{dy_1}{x_1^2+y_1^2+z_1^2}; c = \frac{dz_1}{x_1^2+y_1^2+z_2^2}$$

Substituting the values of a, b and c in (1.81), we get

$$\left(x_1^2 + y_1^2 + z_1^2\right)^3 = 384\lambda^3 x_1^2 y_1^2 z_1^2$$

$\therefore$ The locus of (x_1, y_1, z_1) is $\left(x^2 + y^2 + z^2\right)^3 = 384\lambda^3 xyz$.

Exercise:

1) Find the equation of the plane through the points $(2, 3, -4)$, $(1, -1, 3)$ and parallel to the x-axis.

   Answer: $7y + 4z = 5$

2) Find the equation of the plane passing the point $(-2, 4, 5)$ and parallel to the YZ plane.

   Answer: $x + 2 = 0$

3) Find the equation of the plane passing through the points $(0,1,1)$, $(2, 0, -3)$ and $(3, -2, 0)$.

   Answer: $11x + 10y + 3z = 13$

4) Find the equation of the plane which passes through $(-1, 1, 1)$ and $(1, -1, 1)$ and is perpendicular to the plane $x + 2y + 2z = 5$.

   Answer: $2x + 2y - 3z + 3 = 0$

5) A variable plane is at a constant distance P from the origin and meets the co-ordinate axes in A, B, C. Show that the locus of the centroid of the tetrahedron OABC is $\frac{1}{x^2} + \frac{1}{y^2} + \frac{1}{z^2} = \frac{16}{p^2}$.

6) Find the equation of the plane which is perpendicular to the plane $5x + 3y + 6z + 8 = 0$ and which contains the intersection of planes $x + 2y + 3z - 4 = 0$, $2x + y - z + 5 = 0$.

   Answer: $51x + 51y - 50z + 173 = 0$

7) Show that the points $(0, -1, -1)$, $(4,5,1)$, $(3,9,4)$ and $(-4, 4, 4)$ lie on a plane.

8) Prove that the line joining the points $(2, 2, -1)$, $(3,4,2)$ intersects the line joining the points $(7,0,6)$, $(2,5,1)$.

9) Determine the value of λ and μ for which the two planes $\lambda x + y - 2z + 4 = 0$ and $6x - \mu y - 4z = 0$ are parallel.

Answer: $\lambda = 3, \mu = -2$

10) Determine the value of h for which the planes $3x - 2y + hz = 1$ and $x + hy + 5z = -2$ may be perpendicular to each other.

Answer: $h = -1$

11) Show that the four points $(0,4,3)$, $(-1, -5, -3)$, $(-2, -2, 1)$ and $(1, 1, -1)$ are coplanar.

12) A plane passes through a fixed point (p, q, r) and cuts the axes of coordinates in A, B, C respectively. Show that the locus of the center of sphere OABC is $\frac{p}{x} + \frac{q}{y} + \frac{r}{z} = 2$.

13) Show that the bisector of the obtuse angle between the planes $2x + y - 2z = 4$ and $2x - 3y + 6z + 2 = 0$ is $10x - y + 2z - 11 = 0$.

14) Find the volume of the tetrahedron whose vertices are $(0,1,2)$, $(1,0,2)$, $(1,2,0)$ and $(1,2,1)$.

Answer: $-\frac{1}{3}$

15) Find whether the two points $(2, 3, -5)$ and $(3,4,7)$ lie on the same side or opposite sides of the plane $x + 2y - 2z - 9 = 0$.

Answer: Opposite side

16) Find the plane that bisects the acute angle between the planes $2x - y + 2z + 3 = 0$ and $3x - 2y + 6z + 8 = 0$.

Answer: $23x - 13y + 32z + 45 = 0$

17) Find the angle that the plane $x + 8y - 6z + 16 = 0$ makes with the coordinate planes.

Answer: $\cos^{-1}\frac{1}{\sqrt{101}}$, $\cos^{-1}\frac{3}{\sqrt{101}}$, $\cos^{-1}\frac{6}{\sqrt{101}}$

18) Find the angle between the planes $2x - y + z = 6$ and $x + y + 2z = 7$.

Answer: $\frac{\pi}{3}$

19) Find the distance between the parallel plane
$x - 4y + 8z - 9 = 0$ and $x - 4y + 8z + 18 = 0$.

Answer: 3

20) Prove that the equation $2x^2 - 6y - 12z^2 + 18yz + 2xz + xy = 0$ represents a pair of planes. Find the angle between them.

Answer: $\cos^{-1}\frac{16}{\sqrt{21}}$

21) Show that the equation $6x^2 + 4y^2 - 10z^2 + 3yz + 4zx - 11xy = 0$ represents a pair of planes. Find the angle between them.

Answer: $90°$

22) Find the area of the triangle whose vertices are $(1,2,3)$, $(-2,1,-4)$, $(3,4,-2)$.

Answer: $\frac{\sqrt{1218}}{2}$

23) Find the volume of the tetrahedron whose vertices are $(1,0,0)$, $(0,0,1)$, $(0,0,2)$ and $(1,2,3)$.

Answer: $\frac{1}{3}$

2

Straight Line

2.1 Representation of Line (Introduction)

A straight line may be generated by the intersection of two non-parallel planes. So, the general equation of a straight line is given by the combined equations of two planes as given by

$$a_1 x + b_1 y + c_1 z + d_1 = 0$$

and $a_2 x + b_2 y + c_2 z + d_2 = 0$; provided $\dfrac{a_1}{a_2} \neq \dfrac{b_1}{b_2} \neq \dfrac{c_1}{c_2}$.

Remark:

1) A straight line is also called a right line.
2) The x axis is the intersection of the XZ and XY planes, $y = 0$ and $z = 0$ taken together are its equation. Similarly, $x = 0, z = 0$ are the equations of the y axis, and $x = 0, y = 0$ are the equations of the z axis.

2.2 Equation of a Straight Line in the Symmetrical Form

To find the equations of the line passing through a given point $A(x_1, y_1, z_1)$ and having direction cosines l, m, n:

Let $A(x_1, y_1, z_1)$ be the fixed point on a straight line and $P(x, y, z)$ be any given point on the given line and let $AP = r$.

Let l, m, n be the direction cosines of this line. Assuming projections of AP on the coordinate axes, we get

$$x - x_1 = lr; \; y - y_1 = mr \text{ and } z - z_1 = nr \qquad (2.1)$$

$$\therefore \; \frac{x - x_1}{l} = \frac{y - y_1}{m} = \frac{z - z_1}{n} = r$$

46 Straight Line

which is the equation of a straight line in symmetrical form.
The coordinates (x, y, z) of any points on the line is given by
$x = x_1 + lr \; ; y = y_1 + mr$ and $z = z_1 + nr$
i.e., $(x_1 + lr, y_1 + mr, z_1 + nr) \; ; \; r \in \mathbb{R}$
If $l \neq 0, m \neq 0, n \neq 0$ or equivalently $lmn \neq 0$,

$$\frac{x - x_1}{l} = \frac{y - y_1}{m} = \frac{z - z_1}{n}$$

are the two required equations of the line.

Remark:

1) The equation of the coordinate axes in symmetrical form can be represented respectively by

    $\frac{x-0}{1} = \frac{y-0}{0} = \frac{z-0}{0}; \frac{x-0}{0} = \frac{y-0}{1} = \frac{z-0}{0}; \frac{x-0}{0} = \frac{y-0}{0} = \frac{z-0}{1}$.

2) $(0, 0, 0)$ is the fixed point on the axes and the respective direction cosines of the axes are $(1, 0, 0), (0, 1, 0)$ and $(0, 0, 1)$.

2.3 Equation of a Straight Line Passing through Two Points

To find the equations of the line through two points $A(x_1, y_1, z_1)$ and $B(x_2, y_2, z_2)$:
The direction ratios of the line AB are $x_2 - x_1, y_2 - y_1, z_2 - z_1$
$\therefore$ The required equations of the line AB are

$$\frac{x - x_1}{x_2 - x_1} = \frac{y - y_1}{y_2 - y_1} = \frac{z - z_1}{z_2 - z_1}.$$

1) Find the equation of the lines joining the points $(1, -6, 3)$, and $(0, 2, 8)$.

Sol. Let the equations of the lines joining the points $A(1, -6, 3)$, and $B(0, 2, 8)$ be

$$\frac{x - x_1}{x_2 - x_1} = \frac{y - y_1}{y_2 - y_1} = \frac{z - z_1}{z_2 - z_1}$$

$$\therefore \frac{x - 1}{0 - 1} = \frac{y - (-6)}{2 - (-6)} = \frac{z - 3}{8 - 3}$$

$$\therefore \frac{x - 1}{-1} = \frac{y + 6}{8} = \frac{z - 3}{5}.$$

2.3 Equation of a Straight Line Passing through Two Points

2) Find the equation of the line through the point $(1, 3, 2)$ and parallel to the $\frac{x-5}{3} = \frac{y+4}{-2} = \frac{z-1}{4}$.

Sol. The straight line parallel to

$$\frac{x-5}{3} = \frac{y+4}{-2} = \frac{z-1}{4} \qquad (2.2)$$

is given by

$$\frac{x-x_1}{3} = \frac{y-y_1}{-2} = \frac{z-z_1}{4}.$$

It passes through the point $(1, 3, 2)$

$$\therefore \frac{x-1}{3} = \frac{y-3}{-2} = \frac{z-2}{4}$$

is the required equation of the straight line.

3) Find the equation of a straight line through the point $(8, 9, -10)$ and perpendicular of the two lines $\frac{x-2}{3} = \frac{y-3}{2} = \frac{z+4}{4}$ and $\frac{x+1}{5} = \frac{y-2}{-6} = \frac{z+3}{2}$.

Sol. Let the equation of the line through the point $(8, 9, -10)$ be

$$\frac{x-8}{l} = \frac{y-9}{m} = \frac{z+10}{n}. \qquad (2.3)$$

The line (2.3) is perpendicular to the lines

$$\frac{x-2}{3} = \frac{y-3}{2} = \frac{z+4}{4} \quad \text{and} \quad \frac{x+1}{5} = \frac{y-2}{-6} = \frac{z+3}{2}$$

$$\therefore 3l + 2m + 4n = 0 \text{ and } 5l - 6m + 2n = 0$$

$$\therefore \frac{l}{4+24} = \frac{m}{20-6} = \frac{n}{-18-10}$$

$$\Rightarrow \frac{l}{28} = \frac{m}{14} = \frac{n}{-28}$$

$$\Rightarrow \frac{l}{2} = \frac{m}{1} = \frac{n}{-2}$$

$\therefore$ The required equation of the line is $\frac{x-8}{2} = \frac{y-9}{1} = \frac{z+10}{-2}$.

4) Find the equation of the plane containing the straight line $\frac{x-1}{3} = \frac{y+6}{4} = \frac{z+1}{2}$ and parallel to the straight line $\frac{x-2}{2} = \frac{y-1}{-3} = \frac{z+4}{5}$.

48 Straight Line

Sol. Let the equation of the plane containing the line $\frac{x-1}{3} = \frac{y+6}{4} = \frac{z+1}{2}$ be given by

$$a(x-1) + b(y+6) + c(z+1) = 0 \qquad (2.4)$$

where

$$3a + 4b + 2c = 0. \qquad (2.5)$$

Since the plane (2.4) is parallel to the line

$$\frac{x-2}{2} = \frac{y-1}{-3} = \frac{z+4}{5}$$

$$\Rightarrow 2a - 3b + 5c = 0 \qquad (2.6)$$

By solving Equations (2.5) and (2.6), we get $\frac{a}{26} = \frac{b}{-11} = \frac{c}{-17}$
Substituting the values of $a, b,$ and c in Equation (2.4), we get

$$26(x-1) - 11(y+6) - 17(z+1) = 0$$
$$\therefore 26x - 11y - 17z = 109.$$

5) Find the equation of the straight line passing through the point $(-1, 1, -3)$ and perpendicular to the straight line $\frac{x-3}{-2} = \frac{y+1}{3} = \frac{z-2}{-4}$.

Sol. Any point on the given straight line is

$$\frac{x-3}{-2} = \frac{y+1}{3} = \frac{z-2}{-4} \qquad (2.7)$$

is given by $(-2k+3,\ 3k-1,\ -4k+2)$.

Let this point be the foot of the perpendicular of the line drawn from the pint $(-1,\ 1, -3)$ on the line (2.7).
$\therefore$ The direction ratios of the perpendicular line are given by

$$-2r + 3 + 1;\ 3r - 1 - 1;\ -4r + 2 + 3$$

i.e., $(-2r + 4,\ 3r - 2, -4r + 5)$
The required line is perpendicular to (2.7),

$$\therefore -2(-2r+4) + 3(3r-2) - 4(-4r+5) = 0$$
$$\therefore 4r - 8 + 9r - 6 + 16r - 20 = 0$$
$$\therefore 29r = 34$$
$$\therefore r = \frac{34}{29}$$

∴ The coordinates of the foot of the perpendicular are

$$\left[-2\left(\frac{34}{29}\right)+3,3\left(\frac{34}{29}\right)-1,-4\left(\frac{34}{29}\right)+2\right]=\left(\frac{19}{29},\frac{73}{29},\frac{-78}{29}\right).$$

The equation of the perpendicular line will be the equation of the straight line passing through the point $(-1, 1, -3)$ and $\left(\frac{19}{29}, \frac{73}{29}, \frac{-78}{29}\right)$ is given by

$$\frac{x+1}{\frac{19}{29}+1} = \frac{y-1}{\frac{73}{29}-1} = \frac{z+3}{\frac{-78}{29}+3}$$

$$\therefore \frac{x+1}{48} = \frac{y-1}{44} = \frac{z+3}{9}.$$

2.4 Transformation from the Unsymmetrical to the Symmetrical Form

Let the equation of a straight line be given by

$$a_1x + b_1y + c_1z + d_1 = 0, \tag{2.8}$$

$$a_2x + b_2y + c_2z + d_2 = 0. \tag{2.9}$$

To transform these equations to the symmetrical form we require

1) The direction ratios of the line and
2) The coordinates of any one point on it.

If (l, m, n) be the direction cosines of the line, since this line is perpendicular to both the normals to the planes (2.8) and (2.9), we get

$$a_1 l + b_1 m + c_1 n = 0$$

$$a_2 l + b_2 m + c_2 n = 0$$

$$\Rightarrow \frac{l}{b_1 c_2 - b_2 c_1} = \frac{m}{c_1 a_2 - c_2 a_1} = \frac{n}{a_1 b_2 - a_2 b_1}.$$

Now we require the coordinates of any one point on the line and there is an infinite number of points from which we can choose.

Let us find the point of intersection of the line with the plane $z = 0$. Since this point lies on the planes (2.8) and (2.9), we get

$$a_1x + b_1y + d_1 = 0 \text{ and } a_2x + b_2y + d_2 = 0$$

50 Straight Line

$$\therefore \frac{x}{b_1 d_2 - b_2 d_1} = \frac{y}{d_1 a_2 - d_2 a_1} = \frac{1}{a_1 b_2 - a_2 b_1}.$$

$\therefore$ The coordinates of any fixed point on the line are $\left(\frac{b_1 d_2 - b_2 d_1}{a_1 b_2 - a_2 b_1}, \frac{d_1 a_2 - d_2 a_1}{a_1 b_2 - a_2 b_1}, 0\right)$.

$\therefore$ The equation of the straight line in symmetrical form is

$$\frac{x - \frac{b_1 d_2 - b_2 d_1}{a_1 b_2 - a_2 b_1}}{b_1 c_2 - b_2 c_1} = \frac{y - \frac{d_1 a_2 - d_2 a_1}{a_1 b_2 - a_2 b_1}}{c_1 a_2 - c_2 a_1} = \frac{z - 0}{a_1 b_2 - a_2 b_1}. \qquad (2.10)$$

Conversely, if the equation of any line be given in symmetric form as $\frac{x-a}{l} = \frac{y-b}{m} = \frac{z-c}{n}$ then simplifying the first and second ratios, we get $mx - ma = ly - lb$

$$\text{i.e., } mx - ly = ma - lb \qquad (2.11)$$

and then equating second and third ratios, we get $my - nb = mz - mc$

$$\text{i.e., } ny - mz = nb - mc \qquad (2.12)$$

Equations (2.11) and (2.12) combinedly denotes the equation of the above line in plane form.

6) Find a symmetrical form of the equations of the line $x + y + z + 4 = 0$; $2x + 3y + 4z + 10 = 0$.

Sol. Let l, m, n be the direction ratios of the line
$\therefore l + m + n = 0$ and $2l + 3m + 4n = 0$
i.e., $\frac{l}{(1)(4)-(3)(1)} = \frac{m}{(1)(2)-(4)(1)} = \frac{n}{(1)(3)-(2)(1)}$
i.e., $\frac{l}{1} = \frac{m}{-2} = \frac{n}{1}$
Let the line meets the XY plane at $(\alpha, \beta, 0)$ then we get $\alpha + \beta + 4 = 0$ and $2\alpha + 3\beta + 10 = 0$.
$\therefore \alpha = -2$ and $\beta = -2$
$\therefore$ The equation of the line in symmetric form is $\frac{x+2}{1} = \frac{y+2}{-2} = \frac{z}{1}$.

7) Obtain the symmetrical form of the equations of the line $x - 2y + 3z = 4$ and $2x - 3y + 4z = 5$.

Sol. Let l, m, n be the direction ratios of the line $l - 2m + 3n = 4$ and $2l - 3m + 4n = 5$.
i.e., $\frac{l}{(-2)(4)-(-3)(3)} = \frac{m}{(3)(2)-(1)(4)} = \frac{n}{(1)(-3)-(-2)(2)}$

$$\therefore \frac{l}{1} = \frac{m}{2} = \frac{n}{1}.$$

2.4 Transformation from the Unsymmetrical to the Symmetrical Form

Let the line meets the XY plane at $(\alpha, \beta, 0)$ then we get $\alpha - 2\beta = 4$ and $2\alpha - 3\beta = 5$.

$\therefore \alpha = -2$ and $\beta = -3$

$\therefore$ The equation of the line in symmetric form is $\frac{x+2}{1} = \frac{y+3}{2} = \frac{z-0}{1}$.

8) Find the equations of the line through the point $(1, 2, 4)$ parallel to the line $3x + 2y - z = 4$ and $x - 2y - 2z = 5$.

Sol. Let l, m, n be the direction ratios of the line $3l + 2m - n = 4$ and $l - 2m - 2n = 5$.

$$\therefore \frac{l}{(2)(-2)-(-2)(-1)} = \frac{m}{(-1)(1)-(-2)(3)} = \frac{n}{(3)(-2)-(1)(2)}$$

$$\Rightarrow \frac{l}{-6} = \frac{m}{5} = \frac{n}{-8}$$

$\therefore$ The equation of the line in symmetrical form is $\frac{x-1}{-6} = \frac{y-2}{5} = \frac{z-4}{-8}$.

9) Find the angle between the lines $x - 4y + 3z = 0 = 3x + 2y + 2z$ and $2x + y - z = 0 = 2x - y - 2z$.

Sol. The given lines are

$$\left.\begin{array}{r} x - 4y + 3z = 0 \\ 3x + 2y + 2z = 0 \end{array}\right\}, \quad (2.13)$$

and

$$\left.\begin{array}{r} 2x + y - z = 0 \\ 2x - y - 2z = 0 \end{array}\right\}. \quad (2.14)$$

Let l, m, n be the direction ratios of the line (2.13), then $l - 4m + 3n = 0$ and $3l + 2m + 2n = 0$.

$$\therefore \frac{l}{(-4)(2)-(3)(2)} = \frac{m}{(3)(3)-(2)(1)} = \frac{n}{(2)(1)-(3)(-4)}$$

$$\therefore \frac{l}{-14} = \frac{m}{7} = \frac{n}{14}$$

$$\therefore \frac{l}{-2} = \frac{m}{1} = \frac{n}{2}$$

$\therefore$ Direction ratios of line (2.13) are $-2, 1, 2$.
Similarly, direction ratios of the line (2.14) are $-3, -2, -4$.

Let θ be the angle between the two lines then

$$\cos\theta = \frac{l_1 l_2 + m_1 m_2 + n_1 n_2}{\sqrt{l_1^2 + m_1^2 + n_1^2}\sqrt{l_2^2 + m_2^2 + n_2^2}}$$

$$\therefore \cos\theta = \frac{(-2)(-3) + (1)(-2) + (2)(-4)}{\sqrt{4+1+4}\sqrt{9+4+16}} = 0$$

$\therefore$ The angle between the line is $\frac{\pi}{2}$.

10) Find the angle between the lines

$$3x + 2y + z - 5 = 0 = x + y - 2z - 3$$
$$2x - y - z = 0 = 7x + 10y - 8z.$$

Sol. The given lines are

$$\left.\begin{array}{r}3x + 2y + z = 5 \\ x + y - 2z = 3\end{array}\right\}, \quad (2.15)$$

and

$$\left.\begin{array}{r}2x - y - z = 0 \\ 7x + 10y - 8z = 0\end{array}\right\}. \quad (2.16)$$

Let l_1, m_1, n_1 be the direction ratios of the line (2.15) then

$$\frac{l_1}{(2)(-2) - (1)(1)} = \frac{m_1}{(1)(1) - (3)(-2)} = \frac{n_1}{3(1) - (2)(1)}$$

$$\therefore \frac{l_1}{-5} = \frac{m_1}{7} = \frac{n_1}{1}.$$

$\therefore$ The direction ratios of the line (2.15) are $-5, 7, 1$.

Let l_2, m_2, n_2 be the direction ratios of the line (2.16) then

$$\frac{l_2}{(-1)(-8) - (-1)(10)} = \frac{m_2}{(-1)(7) - (-8)(2)} = \frac{n_2}{(2)(10) - (-1)(7)}$$

$$\therefore \frac{l_2}{18} = \frac{m_2}{9} = \frac{n_2}{27}$$

$$\therefore \frac{l_2}{2} = \frac{m_2}{1} = \frac{n_2}{3}.$$

Let θ be the angle between the two lines then

$$\cos\theta = \frac{l_1 l_2 + m_1 m_2 + n_1 n_2}{\sqrt{l_1^2 + m_1^2 + n_1^2}\sqrt{l_2^2 + m_2^2 + n_2^2}}$$

$$\therefore \cos\theta = \frac{(2)(-5) + (1)(7) + (3)(1)}{\sqrt{25+49+1}\sqrt{4+1+9}} = 0$$

$\therefore$ The angle between the line is $\frac{\pi}{2}$.

11) Show that the condition for the lines $x = az + b$, $y = cz + d$ and $x = a_1 z + b_1$, $y = az + d_1$ to be perpendicular is $aa_1 + cc_1 + 1 = 0$.

Sol. The given lines are

$$\left.\begin{array}{l} x - az = b \\ y - cz = d \end{array}\right\}, \qquad (2.17)$$

and

$$\left.\begin{array}{l} x - a_1 z = b_1 \\ y - c_1 z = d_1 \end{array}\right\}. \qquad (2.18)$$

Let l_1, m_1, n_1 be the direction ratios of the line (2.17) then

$$\frac{l_1}{(0)(-c) - (1)(-a)} = \frac{m_1}{(-a)(0) - (-c)(1)} = \frac{n_1}{(1)(1) - (0)}$$

$$\therefore \frac{l_1}{a} = \frac{m_1}{c} = \frac{n_1}{1}$$

$\therefore$ The direction ratios of the line (2.17) are $a, c, 1$.
Similarly, the direction ratios of the line (2.18) are $a_1, c_1, 1$.
Since the given lines are perpendicular

$$l_1 l_2 + m_1 m_2 + n_1 n_2 = 0$$

$$\therefore aa_1 + cc_1 + 1 = 0.$$

2.5 Angle between a Line and a Plane

To find the angle between the line $\frac{x-x_1}{l} = \frac{y-y_1}{m} = \frac{z-z_1}{n}$
and the plane $ax + by + cz + d = 0$:

The angle between a line and a plane is the complement of the angle between the line and the normal to the plane.

Since the direction cosines of the normal to the given plane and of the given line are proportional to $a, b, c,$ and l, m, n respectively we get

$$\cos(90° - \theta) = \frac{al + bm + cn}{\sqrt{l^2 + m^2 + n^2}\sqrt{a^2 + b^2 + c^2}}$$

where θ is the required angle between the plane and the line.

54 Straight Line

i.e.; $\theta = \sin^{-1}\left[\dfrac{al+bm+cn}{\sqrt{l^2+m^2+n^2}\sqrt{a^2+b^2+c^2}}\right]$.

Remark:

1) If the line is parallel to the plane, then $\theta = 0°$ then $al + bm + cn = 0$.

2) If the line is perpendicular to the plane, then $\dfrac{l}{a} = \dfrac{m}{b} = \dfrac{n}{c}$.

11) Find the angle between the line $\dfrac{x+1}{2} = \dfrac{y}{3} = \dfrac{z-3}{6}$ and the plane $3x + y + z = 7$.

Sol. The equation of the line and the plane are given by
$\dfrac{x+1}{2} = \dfrac{y}{3} = \dfrac{z-3}{6}$ and $3x + y + z = 7$.
Here $(2)(3) + (3)(1) + (6)(1) = 15 \neq 0$.
$\therefore$ The line is not parallel to the plane.
Also, $\dfrac{2}{3} \neq \dfrac{3}{1} \neq \dfrac{6}{1}$
So, the line is not normal to the plane.
The angle between the line and the plane is given by

$$\sin\theta = \dfrac{al+bm+cn}{\sqrt{a^2+b^2+c^2}\sqrt{l^2+m^2+n^2}} = \dfrac{2(3)+(3)(1)+(6)(1)}{\sqrt{4+9+36}\sqrt{9+1+1}}$$

$$\therefore \sin\theta = \dfrac{15}{7\sqrt{11}}$$

$$\therefore \theta = \sin^{-1}\left(\dfrac{15}{7\sqrt{11}}\right).$$

2.6 Point of Intersection of a Line and a Plane

Let the equation of a line and a plane be given by

$$\dfrac{x-x_1}{l} = \dfrac{y-y_1}{m} = \dfrac{z-z_1}{n} (= k), \qquad (2.19)$$

and

$$ax + by + cz + d = 0. \qquad (2.20)$$

The coordinates of any points on the line (2.19) can be written as

$$(x_1 + lk, y_1 + mk, z_1 + nk). \qquad (2.21)$$

If this point lies on the plane (2.20) then

$$a(x_1 + lk) + b(y_1 + mk) + c(z_1 + nk) + d = 0$$

$$\therefore k = \dfrac{-ax_1 + by_1 + cz_1 + d}{al + bm + cn}.$$

Substituting the value of k in (2.21) we get the coordinates of that point where the line meets the plane.

Remark:

If the line (2.19) is parallel to the plane (2.20) then $al + bm + cn = 0$ and hence, we get no finite value of k. So, there are infinite points of intersection.

12) Show that the $\frac{(x-2)}{3} = \frac{y-3}{4} = \frac{z-4}{5}$ is parallel to the plane $2x+y-2z = 3$.

Sol. Here $l = 3, m = 4, n = 5$ and $a = 2, b = 1, c = -2$

$$al + bm + cn = (2)(3) + (1)(4) + 5(-2) = 0.$$

$\therefore$ The line $\frac{x-2}{3} = \frac{y-3}{4} = \frac{z-4}{5}$ is parallel to the plane $2x + y - 2z = 3$.

13) Find the equations of the line through the point $(-2, 3, 4)$ and parallel to the planes $2x + 3y + 4z = 5$ and $3x + 4y + 5z = 6$.

Sol. Let l, m, n be the direction ratios of the planes
$2l + 3m + 4n = 0$ and $3l + 4m + 5n = 0$

$$\therefore \frac{l}{(3)(5) - (4)(4)} = \frac{m}{(4)(3) - (5)(2)} = \frac{n}{(2)(4) - (3)(3)}$$

$$\therefore \frac{l}{-1} = \frac{m}{2} = \frac{n}{-1}$$

$$\therefore \frac{l}{1} = \frac{m}{-2} = \frac{n}{1}$$

$\therefore$ Required equations of the line pass through the point $(-2, 3, 4)$ and parallel to the planes is given by $\frac{x+2}{1} = \frac{y-3}{-2} = \frac{z-4}{1}$.

2.7 Conditions for a Line to Lie in a Plane

To find the conditions for the line

$$\frac{x - x_1}{l} = \frac{y - y_1}{m} = \frac{z - z_1}{n} \qquad (2.22)$$

to lie in the plane

$$ax + by + cz + d = 0. \qquad (2.23)$$

The line (2.22) entirely lies on the plane (2.23) if and only if every point of the line is a point of the plane, i.e., the point $(lk + x_1, mk + y_1, nk + z_1)$ lies on the plane for all values of k.

56 Straight Line

$\therefore$ Equation (2.23) becomes,

$$a(x_1 + lk) + b(y_1 + mk) + c(z_1 + nk) + d = 0$$
$$\therefore (ax_1 + by_1 + cz_1 + d) + k(al + bm + cn) = 0$$
$$\Rightarrow \left.\begin{array}{r} al + bm + cn = 0 \\ ax_1 + by_1 + cz_1 + d = 0 \end{array}\right\}$$

which are the required two conditions.

Remark:

Geometrically these conditions state that a line lies in a given plane if

1) The normal to the plane is perpendicular to the line and
2) Any one point of the line lies in the plane

Corollary:

The general equation of a plane containing the line

$$\frac{x - x_1}{l} = \frac{y - y_1}{m} = \frac{z - z_1}{n} \qquad (2.24)$$

is $A(x - x_1) + B(y - y_1) + C(z - z_1) = 0$ where

$$Al + Bm + Cn = 0 \qquad (2.25)$$

Here A, B, C are parameters subject to the condition (2.25) i.e., The set of planes containing the line (2.24) is

$$\begin{cases} A(x - x_1) + B(y - y_1) + C(z - z_1) = 0, \\ Al + Bm + Cn = 0 \end{cases}$$

2.8 Condition of Coplanarity of Two Straight Lines

The necessary and sufficient condition that the two lines

$$\frac{x - x_1}{l_1} = \frac{y - y_1}{m_1} = \frac{z - z_1}{n_1} \qquad (2.26)$$

and $\quad \dfrac{x - x_2}{l_2} = \dfrac{y - y_2}{m_2} = \dfrac{z - z_2}{n_2} \qquad (2.27)$

2.8 Condition of Coplanarity of Two Straight Lines

will be coplanar is

$$\begin{vmatrix} x_2 - x_1 & y_2 - y_1 & z_2 - z_1 \\ l_1 & m_1 & n_1 \\ l_2 & m_2 & n_2 \end{vmatrix} = 0. \qquad (2.28)$$

Necessary condition:

Let the two given lines be coplanar and the equation of any plane containing the line (2.26) is

$$A(x - x_1) + B(y - y_1) + c(z - z_1) = 0 \qquad (2.29)$$

where A, B, C are not all zero satisfying the condition

$$Al_1 + Bm_1 + Cn_1 = 0 \qquad (2.30)$$

The plane (2.29) will contain the line (2.27) if
(a) the point (x_2, y_2, z_2) lies on it

$$\text{i.e., } A(x_2 - x_1) + B(y_2 - y_1) + C(z_2 - z_1) = 0 \qquad (2.31)$$

(b) the line is perpendicular to the normal to the plane

$$\text{i.e., } Al_2 + Bm_2 + Cn_2 = 0. \qquad (2.32)$$

Two lines will be coplanar if the three linear homogeneous Equations (2.30), (2.31), (2.32) in A, B, C are consistent so that

$$\begin{vmatrix} x_2 - x_1 & y_2 - y_1 & z_2 - z_1 \\ l_1 & m_1 & n_1 \\ l_2 & m_2 & n_2 \end{vmatrix} = 0$$

which is the required condition for the lines to intersect.

Let this condition be satisfied then the required equation of the plane containing the two lines is

$$\begin{vmatrix} x_2 - x_1 & y_2 - y_1 & z_2 - z_1 \\ l_1 & m_1 & n_1 \\ l_2 & m_2 & n_2 \end{vmatrix} = 0.$$

58 Straight Line

Sufficient condition:

Two lines are coplanar if and if they intersect or are parallel.

Let us first consider the case of intersection.

The general coordinates of the points on the lines (2.26) and (2.27) respectively for all values of k_1 and k_2 is given by

$$(l_1 k_1 + x_1, m_1 k_1 + y_1, n_1 k_1 + z_1),$$

and

$$(l_2 k_2 + x_2, m_2 k_2 + y_2, n_2 k_2 + z_2).$$

The lines (2.26) and (2.27) intersect so these points should coincide for the same values of k_1 and k_2, we get

$$(x_1 - x_2) + l_1 k_1 - l_2 k_2 = 0$$
$$(y_1 - y_2) + m_1 k_1 - m_2 k_2 = 0$$
$$(z_1 - z_2) + n_1 k_1 - n_2 k_2 = 0$$

$$\Leftrightarrow \begin{vmatrix} x_1 - x_2 & l_1 & l_2 \\ y_1 - y_2 & m_1 & m_2 \\ z_1 - z_2 & n_1 & n_2 \end{vmatrix} = 0$$

$$\Leftrightarrow \begin{vmatrix} x_2 - x_1 & y_2 - y_1 & z_2 - z_1 \\ l_1 & m_1 & n_1 \\ l_2 & m_2 & n_2 \end{vmatrix} = 0.$$

which represents the plane.

$\therefore$ The lines (2.26) and (2.27) are coplanar.

This condition is satisfied if lines are parallel.

Remark:

1) In general, the equation

$$\begin{vmatrix} x - x_1 & y - y_1 & z - z_1 \\ l_1 & m_1 & n_1 \\ l_2 & m_2 & n_2 \end{vmatrix} = 0$$

represents the plane that passes through the line (2.26) and is parallel to the line (2.27), and the equation.

$$\begin{vmatrix} x - x_2 & y - y_2 & z - z_2 \\ l_1 & m_1 & n_1 \\ l_2 & m_2 & n_2 \end{vmatrix} = 0$$

2.8 Condition of Coplanarity of Two Straight Lines

represents the plane that passes through the line (2.27) and is parallel to the line (2.26).

In case the lines are coplanar then the point (x_2, y_2, z_2) lies on the first plane and the point (x_1, y_1, z_1) on the second. Moreover, these two equations are then identical.

$\therefore$ The plane containing two coplanar lines is the one that passes through one line and is parallel to the other or through one line and any point on the other.

2) Two lines will intersect if and if, there exists a point whose coordinates satisfy the four equations, two of each line so that for intersection, we require that the four linear equations in three unknowns should be consistent.

3) The condition for the lines whose equations, given in the unsymmetrical form are

$$a_1 x + b_1 y + c_1 z + d_1 = 0 = a_2 x + b_2 y + c_2 z + d_2$$

and $a_3 x + b_3 y + c_3 z + d_3 = 0 = a_4 x + b_4 y + c_4 z + d_4$
to intersect is the condition for the consistency of these four equations.

i.e., $\begin{vmatrix} a_1 & b_1 & c_1 & d_1 \\ a_2 & b_2 & c_2 & d_2 \\ a_3 & b_3 & c_3 & d_3 \\ a_4 & b_4 & c_4 & d_4 \end{vmatrix} = 0.$

14) Find the equation of the plane containing the line $\frac{x+2}{2} = \frac{y+3}{3} = \frac{z-4}{-2}$ and the point $(0, 6, 0)$.

Sol. The general equation of the plane containing the given line
$\frac{x+2}{2} = \frac{y+3}{3} = \frac{z-4}{-2}$ is

$$A(x+2) + B(y+3) + C(z-4) = 0 \qquad (2.33)$$

where A, B, C are parameters subjected to the condition $Al + Bm + Cn = 0$
i.e.,

$$2A + 3B - 2C = 0 \qquad (2.34)$$

The plane (2.33) passes through the point $(0, 6, 0)$, we get

$$2A + 9B - 4C = 0 \qquad (2.35)$$

Eliminating A, B, C from (2.33), (2.34), and (2.35), we get

$$\begin{vmatrix} x+2 & y+3 & z-4 \\ 2 & 3 & -2 \\ 2 & 9 & -4 \end{vmatrix} = 0$$

$\therefore -6(x+2) + 4(y+3) + 12(z-4) = 0$
$\therefore 3x + 2y + 6z - 12 = 0$

is the required equation of the plane.

15) Show that the line $x + 10 = \frac{8-y}{2} = z$ lies in the plane $x + 2y + 3z = 6$.

Sol. The line $\frac{x+10}{1} = \frac{y-8}{-2} = \frac{z-0}{1} = k$ lies in the plane $x + 2y + 3z = 6$ if and only if every point of the line is a point of the plane.
 i.e., $(k - 10, -2k + 8, k)$ lie on the planes for all values of k.
 i.e., $k - 10 + 2(-2k + 8) + 3k - 6 = 0$.
$\therefore$ The given line lies in the plane $x + 2y + 3z = 6$.

16) Prove that the plane through (α, β, γ), and the line $x = py + q = rz + s$ is given by

$$\begin{vmatrix} x & py+q & rz+s \\ \alpha & p\beta+q & r\gamma+s \\ 1 & 1 & 1 \end{vmatrix} = 0.$$

Sol. Given line q can be written as

$$\frac{x}{1} = \frac{y + \frac{q}{p}}{\frac{1}{p}} = \frac{z + \frac{s}{r}}{\frac{1}{r}}. \tag{2.36}$$

Let the equation of any plane be

$$Ax + By + Cz + D = 0 \tag{2.37}$$

It will pass through line (2.37), if

$$A(0) + B\left(\frac{-q}{p}\right) + C\left(\frac{-s}{r}\right) + D = 0, \tag{2.38}$$

and

$$A(1) + B\left(\frac{1}{p}\right) + C\left(\frac{1}{r}\right) = 0. \tag{2.39}$$

The plane will pass through (α, β, γ) if

$$A\alpha + B\beta + C\gamma + D = 0. \tag{2.40}$$

2.8 Condition of Coplanarity of Two Straight Lines

Subtracting (2.38) from (2.39) and (2.40), we get

$$Ax + B\left(y + \frac{q}{p}\right) + C\left(z + \frac{s}{r}\right) = 0, \qquad (2.41)$$

$$A\alpha + B\left(\beta + \frac{q}{p}\right) + C\left(\gamma + \frac{s}{r}\right) = 0. \qquad (2.42)$$

Eliminating A, B, C from (2.39), (2.41), and (2.42), we get

$$\begin{vmatrix} x & y + \frac{q}{p} & z + \frac{s}{r} \\ \alpha & \gamma + \frac{q}{p} & \gamma + \frac{s}{r} \\ 1 & \frac{1}{p} & \frac{1}{r} \end{vmatrix} = 0$$

$$\Rightarrow \begin{vmatrix} x & py + q & zr + s \\ \alpha & p\beta + q & r\gamma + s \\ 1 & 1 & 1 \end{vmatrix} = 0.$$

17) Show that the line $\frac{x+1}{-2} = \frac{y+2}{3} = \frac{z+5}{4}$ lies on the plane $x + 2y - z = 0$.

Sol. The given line is

$$\frac{x+1}{-2} = \frac{y+2}{3} = \frac{z+5}{4} \qquad (2.43)$$

and the given plane is

$$x + 2y - z = 0 \qquad (2.44)$$

The line (2.43) lies in the plane (2.44) if

(i) $al + bm + cn = 0$.

(ii) One point $(-1, -2, -5)$ on the line lies on the plane.

Direction ratios l, m, n of a line are $-2, 3, 4$ respectively.

$$\therefore al + bm + cn = (1)(-2) + 2(3) + (-1)(4) = 0$$

Moreover; the point $(-1, -2, -5)$ lies on the plane (2.45) if $-1 + 2(-2) - (-5) = 0$.

$\therefore$ The line (2.43) lies in the plane (2.44).

18) Find the equation of the plane containing the line $\frac{x+1}{-3} = \frac{y-3}{2} = \frac{z+2}{1}$ and the point $(0, 7, -7)$ and show that the line $\frac{x}{1} = \frac{y-7}{-3} = \frac{z+7}{2}$ also lies in the same plane.

Sol. The given line is

$$\frac{x+1}{-3} = \frac{y-3}{2} = \frac{z+2}{1}. \tag{2.45}$$

Now any plane containing the line (2.45) is

$$A(x+1) + B(y-3) + C(z+2) = 0 \tag{2.46}$$

$$\text{where } -3A + 2B + C = 0 \tag{2.47}$$

Since the plane passes through the point $(0, 7, -7)$, we get

$$A + 4B - 5C = 0 \tag{2.48}$$

Solving (2.47) and (2.48), we get

$$\frac{A}{-10-4} = \frac{B}{1-15} = \frac{C}{-12-2} \Rightarrow \frac{A}{1} = \frac{B}{1} = \frac{C}{1}$$

Substituting these values of A, B, and C in (2.46), we get

$$1(x+1) + 1(y+3) + 1(z+2) = 0$$
$$\therefore x + y + z = 0 \tag{2.49}$$

is the required equation.

Let the line

$$\frac{x}{1} = \frac{y-7}{-3} = \frac{z+7}{2} \tag{2.50}$$

lies on the plane (2.49) if

(a) one point $(0, 7, -7)$ lies on the plane (2.49)

(b) $al + bm + cn = 0$

Now point $(0, 7, -7)$ lies on the plane (2.49) is $0 + 7 - 7 = 0$

$$al + bm + cn = (1)(1) + (1)(-3) + (1)(2) = 0.$$

$\therefore$ The line (2.50) lies in the plane (2.49).

19) Find the equation to the plane which passes through the z axis and is perpendicular to the line $\frac{x-1}{\cos \alpha} = \frac{y+2}{\sin \alpha} = \frac{z-3}{0}$.

Sol. The equations of z axis are

$$x = 0, y = 0. \tag{2.51}$$

2.8 Condition of Coplanarity of Two Straight Lines

Any plane through (2.51) is

$$\lambda x + \lambda y = 0. \tag{2.52}$$

If (2.52) is perpendicular to the line $\frac{x-1}{\cos\alpha} = \frac{y+2}{\sin\alpha} = \frac{z-3}{0}$
then $\frac{1}{\cos\alpha} = \frac{\alpha}{\sin\alpha} = \frac{0}{0}. \Rightarrow \lambda = \tan\alpha$
$\therefore$ Equation (2.52) becomes,

$$x + (\tan\alpha)\, y = 0$$

$$\therefore x\cos\alpha + y\sin\alpha = 0.$$

20) Prove that the lines $\frac{x+1}{3} = \frac{y+3}{5} = \frac{z+5}{7}$ and $\frac{x-2}{1} = \frac{y-4}{3} = \frac{z-6}{5}$ intersect. Find their point of intersection and the plane in which they lie.

Sol. The given lines are

$$\frac{x+1}{3} = \frac{y+3}{5} = \frac{z+5}{7} \tag{2.53}$$

and

$$\frac{x-2}{1} = \frac{y-4}{3} = \frac{z-6}{5}. \tag{2.54}$$

The equation of the plane which contains the line (2.53) and is parallel to (2.54) is given by

$$\begin{vmatrix} x+1 & y+3 & z+5 \\ 3 & 5 & 7 \\ 1 & 3 & 5 \end{vmatrix} = 0$$

$$\therefore (x+1)(25-21) - (y+3)(15-7) + (z+5)(9-5) = 0$$

$$\therefore 4x - 8y + 4z = 0$$

$$\therefore x - 2y + z = 0. \tag{2.55}$$

The plane (2.55) passes through $(2, 4, 6)$ a point on the line (2.54) if $2 - 2(4) + 6 = 0$ or if $0 = 0$ which is true.
$\therefore$ Given two lines are coplanar and lie in the plane $x - 2y + z = 0$.

21) Show that the lines $\frac{x+1}{-3} = \frac{y-3}{2} = \frac{z+2}{1}$ and $\frac{x}{1} = \frac{y-7}{-3} = \frac{z+7}{2}$ intersect. Find the coordinates of the point of intersection and the equation to the plane containing them.

Sol. The given lines are

$$\frac{x+1}{-3} = \frac{y-3}{2} = \frac{z+2}{1} = k, \tag{2.56}$$

and
$$\frac{x}{1} = \frac{y-7}{-3} = \frac{z-7}{2}. \quad (2.57)$$

Any point on the line (2.56) is given by
$$(-3k-1, 2k+3, k-2) \quad (2.58)$$

It lies on the line (2.57) if
$$\frac{-3k-1}{1} = \frac{2k-4}{-3} = \frac{k+5}{2} \quad (2.59)$$

From (2.56) and (2.57),
$$9k+3 = 2k-4 \quad \Rightarrow k = -1.$$

Substituting the value of k in (2.59), we get
$$\frac{3-1}{1} = \frac{-2-4}{-3} = \frac{-1+5}{2}$$

$\Rightarrow 2 = 2 = 2$ which is true.

∴ Two lines intersect and from (2.58), the point of intersection is $(-3(-1)-1, 2(-1)+3, -1-2) = (2, 1, -3)$.

Moreover, the plane containing the lines (2.56) and (2.57) is
$$\begin{vmatrix} x+1 & y-3 & z+2 \\ -3 & 2 & 1 \\ 1 & -3 & 2 \end{vmatrix} = 0$$

$$\therefore (x+1)(4+3) - (y-3)(-6-1) + (z+2)(+9-2) = 0$$

$$\therefore 7x + 7y + 7z = 0$$

$$\therefore x + y + z = 0$$

is the required equation of the plane containing the given two lines.

22) Prove that the two lines $x = az+b, y = cz+d$ and $x = \alpha z+\beta, y = \gamma z+\delta$ intersect if $(\beta - b)(\gamma - c) = (\delta - d)(\alpha - a)$.

Sol. The first line is $x = az+b, y = cz+d$

$$\Rightarrow \frac{x-b}{a} = z \text{ and } \frac{y-d}{c} = z$$

2.8 Condition of Coplanarity of Two Straight Lines

$$\Rightarrow \frac{x-b}{a} = \frac{y-d}{c} = z. \qquad (2.60)$$

The given second line is $x = \alpha z + \beta, y = \gamma z + \delta$

$$\Rightarrow \frac{x-\beta}{\alpha} = z, \quad \frac{y-\delta}{\gamma} = z$$

$$\Rightarrow \frac{x-\beta}{\alpha} = \frac{y-\delta}{\gamma} = z. \qquad (2.61)$$

The equation of the plane through the line (2.60) and parallel to (2.61) is given by

$$\begin{vmatrix} x-b & y-d & z \\ \alpha & \gamma & 1 \\ a & c & 1 \end{vmatrix} = 0. \qquad (2.62)$$

The two lines will be coplanar if the point $(\beta, \delta, 0)$ on the second line lies on the plane (2.62).
∴ The required condition is

$$\begin{vmatrix} \beta-b & \delta-d & 0 \\ \alpha & \gamma & 1 \\ a & c & 1 \end{vmatrix} = 0$$

$$\therefore (\beta - b)(\gamma - c) - (\delta - d)(\alpha - a) = 0.$$

23) Prove that the lines $\frac{x+1}{1} = \frac{y+1}{2} = \frac{z+1}{3}$ and $x + 2y + 3z - 8 = 0 = 2x + 3y + 4z - 11$ are coplanar and find the coordinates of the point of intersection find also the equation of the plane containing them.

Sol. The given lines are

$$\frac{x+1}{1} = \frac{y+1}{2} = \frac{z+1}{3}, \qquad (2.63)$$

and

$$\left. \begin{array}{l} x + 2y + 3z - 8 = 0 \\ 2x + 3y + 4z - 11 = 0 \end{array} \right\}. \qquad (2.64)$$

Any plane through the line (2.64) is

$$x + 2y + 3z - 8 + k(2x + 3y + 4z - 11) = 0. \qquad (2.65)$$

The line (2.63) lies on this plane if

1) The point $(-1, -1, -1)$ on the line (2.63) lies on this plane.

2) $al + bm + cn = 0$.

Now the point $(-1, -1, -1)$ lies on the plane (2.63) if

$$-1 - 2 - 3 - 8 + k(-2 - 3 - 4 - 11) = 0$$

$$\therefore k = \frac{-7}{10}.$$

Substituting the value of k in (2.65), we get

$$x + 2y + 3z - 8 - \frac{7}{10}(2x + 3y + 4z - 11) = 0$$

$$\therefore 10x + 20y + 30z - 80 - 14x - 21y - 28z + 77 = 0$$

$$\therefore -4x - y + 2z - 3 = 0$$

$$\therefore 4x + y - 2z + 3 = 0.$$

Now, $al + bm + cn = 4(1) + 1(2) + (-2)(3) = 0$.
$\therefore$ The two lines are coplanar and the equation of the plane in which they lie is $4x + y - 2z + 3 = 0$.

The two lines (2.63) and (2.64) will intersect at a point where the line (2.63) meets one of the planes of the second line.

Now any point on the first line is

$$(r - 1, \ 2r - 1, \ 3r - 1) \qquad (2.66)$$

If this lies on the first plane of the 2nd line,

$$r - 1 + 2(2r - 1) + 3(3r - 1) - 8 = 0$$

$$\therefore r = 1.$$

$\therefore$ Substituting the value of r in (2.66), we get $(0, 1, 2)$ as the point of intersection of two lines.

24) A, A^1, B, B^1, C, C^1 are points on the axes, shows that the lines of intersection of the plane $A^1BC, AB^1C^1, B^1CA, BC^1A^1, C^1AB, CA^1B$ are coplanar.

Sol. Let $A(a, 0, 0)$, $A^1(a^1, 0, 0)$, $B(0, b, 0)$, $B^1(0, b^1, 0)$, $C(0, 0, c)$ and $C^1(0, 0, c^1)$ be the given point.

2.8 Condition of Coplanarity of Two Straight Lines

The equation of the plane A^1BC is

$$\frac{x}{a^1} + \frac{y}{b} + \frac{z}{c} = 1 \qquad (2.67)$$

and the equation of the plane AB^1C^1 is

$$\frac{x}{a} + \frac{y}{b^1} + \frac{z}{c^1} = 1 \qquad (2.68)$$

Adding (2.67) and (2.68), we get

$$x\left(\frac{1}{a} + \frac{1}{a^1}\right) + y\left(\frac{1}{b} + \frac{1}{b^1}\right) + z\left(\frac{1}{c} + \frac{1}{c^1}\right) = 2 \qquad (2.69)$$

Thus, the line of intersection of the planes (2.67) and (2.68) lies on the plane (2.69).

By symmetry, the line of intersection of the planes B^1CA, BC^1A^1 and C^1AB^1, C^1AB also lie on the plane (2.69).

∴ All the lines of the given planes are coplanar.

25) Show that the line of intersection of the first two planes is coplanar with the line of intersection of the later two $x+2y-z-3=0$, $3x-y+2z-1=0$, $2x-2y+3z-2=0$, $x-y+z+1=0$ are four planes. and find the equation of the plane containing the two lines.

Sol. The line of intersection of the first two planes is

$$\left.\begin{array}{l} x+2y-z-3=0 \\ 3x-y+2z-1=0 \end{array}\right\} \qquad (2.70)$$

and the line of intersection of the last two planes is

$$\left.\begin{array}{l} 2x-2y+3z-2=0 \\ x-y+z+1=0 \end{array}\right\} \qquad (2.71)$$

Now to prove that the two lines (2.70) and (2.71) are coplanar we reduce the line (2.70) in symmetrical form.

To find direction ratios of the line (2.70), we omit constant,

i.e., $x + 2y - z = 0$

$3x - y + 2z = 0$

$$\therefore \frac{x}{(4-1)} = \frac{y}{(-3-2)} = \frac{z}{(-1-6)}$$

$$\therefore \frac{x}{3} = \frac{y}{-5} = \frac{z}{-7}$$

$\therefore$ The direction ratios of the line are $3, -5, -7$.
For one point on the line (2.70), Let $z = 0$ in (2.70), we get

$$x - 2y - 3 = 0$$
$$3x - y - 1 = 0$$
$$\therefore \frac{x}{-5} = \frac{y}{-8} = \frac{1}{-7}$$
$$\therefore x = \frac{5}{7},\ y = \frac{8}{7},\ z = 0$$

$\therefore$ One point on the line (2.70) is $\left(\frac{5}{7}, \frac{8}{7}, 0\right)$
Hence the equations of the line (2.70) in the symmetrical form are

$$\frac{x - \frac{5}{7}}{3} = \frac{y - \frac{8}{7}}{-5} = \frac{z - 0}{-7}. \tag{2.72}$$

Now any plane through the line (2.71) is

$$2x - 2y + 3z - 2 + k(x - y + z + 1) = 0. \tag{2.73}$$

This plane contains the line (2.72) if

(i) $al + bm + cn = 0$

(ii) One point $\left(\frac{5}{7}, \frac{8}{7}, 0\right)$ on the line (2.72) line on this plane.

Now coefficients x, y, z in the plane (2.73) are $2 + k, -2 - k, 3 + k$ and direction ratios of the line (2.73) are $3, -5, -7$.
$\therefore al + bm + cn = 0$ gives

$$3(2 + k) + (-5)(-2 - k) + (-7)(3 + k) = 0$$
$$\therefore k - 5 = 0$$
$$\therefore k = 5.$$

Substituting $k = 5$ in (2.73), we get

$$2x - 2y + 3z - 2 + 5(x - y + z + 1) = 0$$
$$\therefore 7x - 7y + 8z + 3 = 0. \tag{2.74}$$

The point $\left(\frac{5}{7}, \frac{8}{7}, 0\right)$ lies on (2.74) if $5 - 8 + 0 + 3 = 0$
or $0 = 0$ which is true.
$\therefore$ The two lines are coplanar and they lie in the plane $7x - 7y + 8z + 3 = 0$.

2.9 Skew Lines and the Shortest Distance between Two Lines

Skew lines:
Two straight lines are said to be skewed if they are neither parallel nor intersecting.
    i.e., the lines which do not lie in a plane.

Shortest distance between two lines:

To show that the length of the line intercepted between two lines which are perpendicular to both is the shortest distance between them.
    Let AB and CD be two skew lines given by

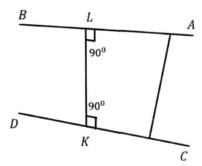

$$\frac{x - x_1}{l_1} = \frac{y - y_1}{m_1} = \frac{z - z_1}{n_1} \qquad (2.75)$$

and
$$\frac{x - x_2}{l_2} = \frac{y - y_2}{m_2} = \frac{z - z_1}{n_2}. \qquad (2.76)$$

Let LK be the line of shortest distance. If (l, m, n) be the direction of LK then LK is perpendicular to both AB and CD so,

$$ll_1 + mm_1 + nn_1 = 0,$$

and $ll_2 + mm_2 + nn_2 = 0$

$$\Rightarrow \frac{l}{m_1 n_2 - m_2 n_1} = \frac{m}{n_1 l_2 - n_2 l_1} = \frac{n}{l_1 m_2 - l_2 m_1}$$

$$\Rightarrow l = \frac{m_1 n_2 - m_2 n_1}{\sqrt{\sum (m_1 n_2 - m_2 n_1)^2}}, \quad m = \frac{n_1 l_2 - n_2 l_1}{\sqrt{\sum (m_1 n_2 - m_2 n_1)^2}}$$

and $n = \dfrac{l_1 m_2 - l_2 m_1}{\sqrt{\sum (m_1 n_2 - m_2 n_1)^2}}.$ \hfill (2.77)

Let the coordinates of A and C are (x_1, y_1, z_1) and (x_2, y_2, z_2) respectively then

$$\begin{aligned} LK &= \text{Length of the shortest distance} \\ &= \text{Projection of } AC \text{ on } PQ \\ &= (x_2 - x_1)\, l + (y_2 - y_1)\, m + (z_2 - z_1)\, n \\ &= \frac{\begin{vmatrix} x_2 - x_1 & y_2 - y_1 & z_2 - z_1 \\ l_1 & m_1 & n_1 \\ l_2 & m_2 & n_2 \end{vmatrix}}{\sqrt{\sum (m_1 n_2 - m_2 n_1)^2}} \end{aligned}$$

To find the equation of the line of shortest distance, we observe that it is coplanar with both the given lines.

$\therefore$ The equation of the plane containing LK and AB will be

$$\begin{vmatrix} x - x_1 & y - y_1 & z - z_1 \\ l_1 & m_1 & n_1 \\ l & m & n \end{vmatrix} = 0, \hfill (2.78)$$

and that of the plane containing the coplanar lines LK and AB will be

$$\begin{vmatrix} x - x_2 & y - y_2 & z - z_2 \\ l_2 & m_2 & n_2 \\ l & m & n \end{vmatrix} = 0. \hfill (2.79)$$

Since LK is the line of intersection of the planes (2.78) and (2.79), so the equation of the line of shortest distance will be

$$\begin{vmatrix} x - x_1 & y - y_1 & z - z_1 \\ l_1 & m_1 & n_1 \\ l & m & n \end{vmatrix} = 0 = \begin{vmatrix} x - x_2 & y - y_2 & z - z_2 \\ l_2 & m_2 & n_2 \\ l & m & n \end{vmatrix}$$

where l, m, n are given by (2.77).

Corollary:

If the lines (2.75) and (2.76) are coplanar, then the length of the shortest distance is zero.

∴ The condition (2.77) becomes

$$\begin{vmatrix} x_2 - x_1 & y_2 - y_1 & z_2 - z_1 \\ l_1 & m_1 & n_1 \\ l_2 & m_2 & n_2 \end{vmatrix} = 0.$$

Alternative Method:

We now find the length of shortest distance and the equation of shortest distance in another way which is very useful to solve numerical problems.

Let the equations of two skew lines AB and CD be

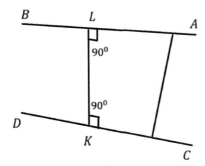

$$\frac{x - x_1}{l_1} = \frac{y - y_1}{m_1} = \frac{z - z_1}{n_1} = (k_1) \quad (2.80)$$

and

$$\frac{x - x_2}{l_2} = \frac{y - y_2}{m_2} = \frac{z - z_2}{n_2} = (k_2) \quad (2.81)$$

Let LK be the line of shortest distance. Now the coordinates of L and K may be taken as $(x_1 + l_1 k_1, y_1 + m_1 k_1, z_1 + n_1 k_1)$, and $(x_2 + l_2 k_2, y_2 + m_2 k_2, z_2 + n_2 k_2)$ respectively.

∴ The direction ratios of LK will be $(x_2 + l_2 k_2 - x_1 - l_1 k_1, y_2 + m_2 k_2 - y_1 - m_1 k_1, z_2 + n_2 k_2 - z_1 - m_1 k_1)$.

Now since LK is perpendicular to both (2.80) and (2.81), we get

$$l_1 (x_2 + l_2 k_2 - x_1 - l_1 k_1) + m_1 (y_2 + m_2 k_2 - y_1 - m_1 k_1) + n_1 (z_2 + m_2 k_2 - z_1 - m_1 k_1) = 0 \quad (2.82)$$

and

$$l_2(x_2 + l_2k_2 - x_1 - l_1k_1) + m_2(y_2 + m_2k_2 - y_1 - m_1k_1) + \\ n_2(z_2 + m_2k_2 - z_1 - m_1k_1) = 0. \quad (2.83)$$

Solving (2.82) and (2.83), we get the values of k_1 and k_2.

Substituting k_1 and k_2 we get coordinates of L and K. Let coordinates of L and K be $(\alpha_1, \beta_1, \gamma_1)$ and $(\alpha_2, \beta_2, \gamma_2)$ respectively then

LK = Length of the shortest distance

$$= \sqrt{(\alpha_2 - \alpha_1)^2 + (\beta_2 - \beta_1)^2 + (\gamma_2 - \gamma_1)^2}.$$

$\therefore$ The equation of the line of shortest distance will be

$$\frac{x - \alpha_1}{\alpha_2 - \alpha_1} = \frac{y - \beta_1}{\beta_2 - \beta_1} = \frac{z - \gamma_1}{\gamma_2 - \gamma_1}.$$

2.10 Equation of Two Skew Lines in Symmetric Form

The equation of two skew lines can be written in the form $y = mx, z = c$, and $y = -mx, z = -c$ by a special choice of axes; where $2c$ is the length of the shortest distance between two lines.

Let AB and CD be two skew lines and LK be their line of shortest distance and also let the length of shortest distance be $2c$.

The Special choice of axes:

Let us consider the middle point LK as the origin O and this line of shortest distance as z - axis.

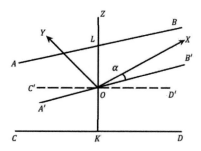

Through O let us draw two lines A^1OB^1 and C^1OD^1 is perpendicular to LK. In this plane, let us take the internal and external bisectors of the angle B^1OD^1 as $x - axis$ and $y - axis$ respectively.

2.10 Equation of Two Skew Lines in Symmetric Form

Let $\angle B^1 OD^1 = 2\alpha$.
$\therefore \{OX, OY, OZ\}$ is a new set of rectangular axes by special choice.
Now the line OB^1 or AB makes angles $\alpha, \frac{\pi}{2} - \alpha, \frac{\pi}{2}$ with the coordinate axes OX, OY and OZ respectively.
Similarly, OD^1 or CD makes angles $-\alpha, \frac{\pi}{2} + \alpha, \frac{\pi}{2}$ with respective axes.
$\therefore$ The direction cosines of AB and CD will be
$\{\cos\alpha, \cos\left(\frac{\pi}{2} - \alpha\right), \cos\frac{\pi}{2}\}$ and $\{\cos(-\alpha), \cos\left(\frac{\pi}{2} + \alpha\right), \cos\frac{\pi}{2}\}$
i.e., $(\cos\alpha, \sin\alpha, 0)$ and $(\cos\alpha, -\sin\alpha, 0)$ respectively.
Now $OL = OK = c (\because LK = 2c)$ and since LK is taken as $z - axis$, so the coordinates of the point L and K will be $(0, 0, c)$ and $(0, 0, -c)$ respectively.

Since the line AB is a straight line passing through $(0, 0, c)$ and having direction cosines $(\cos\alpha, \sin\alpha, 0)$, so its equation will be

$$\frac{x-0}{\cos\alpha} = \frac{y-0}{\sin\alpha} = \frac{z-c}{0} \qquad (2.84)$$

and similarly, the equation of the other line will be

$$\frac{x-0}{\cos\alpha} = \frac{y-0}{-\sin\alpha} = \frac{z+c}{0}. \qquad (2.85)$$

The Equations (2.84) and (2.85) can also be written as

$$y = x\tan\alpha, \ z = c \qquad (2.86)$$

and

$$y = -x\tan\alpha, \ z = -c. \qquad (2.87)$$

Let us assume $m = \tan\alpha$, so Equations (2.86) and (2.87) becomes

$$y = mx, \ z = c \qquad (2.88)$$

and

$$y = -mx, \ z = -c. \qquad (2.89)$$

Equations (2.88) and (2.89) can also be expressed in compact form as

$$y = \pm mx, \ z = \pm c.$$

Remark:

In parametric form, any point on (2.86) and (2.87) can be written as $(k, \tan\alpha, c)$ and $(k_1, -k_1\tan\alpha, -c)$.

74 Straight Line

Alternative Methods:

We can also determine the length and equations of the shortest distance between two given lines by the following methods:

$$\frac{x-x_1}{l_1} = \frac{y-y_1}{m_1} = \frac{z-z_1}{n_1}$$

$$\frac{x-x_2}{l_2} = \frac{y-y_2}{m_2} = \frac{z-z_2}{n_2}.$$

I) Method of Intersection:

The general coordinates of points on the two lines are

$$P(x_1+l_1k, y_1+m_1k, z_1+n_1k), \text{ and } Q(x_2+l_2k_1, y_2+m_2k_1, z_2+n_2k_1) \quad (2.90)$$

The direction ratios of PQ are $x_1-x_2+l_1k-l_2k_1, y_1-y_2+m_1k-m_2k_1, z_1+n_1k-n_2k_1$.

Since PQ is perpendicular to AB and CD both, we get two equations which may be solved for k, k_1. Substituting the values of k and k_1 in equation (2.90) we get the coordinates of the points P and Q.

We can find the shortest distance by distance formula.
i.e., The length of the shortest distance = PQ

$$\frac{x-x_1}{x_2-x_1} = \frac{y-y_1}{y_2-y_1} = \frac{z-z_1}{z_2-z_1}.$$

II) Method of parallel plane:

The length of the shortest distance can be obtained by using the fact that it is equal to the perpendicular from any point on the first line upon the plane drown through the second line parallel to the first.

Remark:

1) If only the shortest distance and its equations are required and the two given lines are in the symmetrical form then the method of projection should be used.

2) If points of intersection of the line of shortest distance with the given lines are also required, then the method of intersection should be used.

2.10 Equation of Two Skew Lines in Symmetric Form

3) If one or both the lines are given in the general form and only the length of the shortest distance is required then the method of the parallel plane should be preferred.

26) Find the length and the equation of the shortest distance between the lines
$\frac{x-3}{2} = \frac{y+15}{-7} = \frac{z-9}{5}; \frac{x+1}{2} = \frac{y-1}{1} = \frac{z-9}{-3}.$

Sol. The given lines are

$$\frac{x-3}{2} = \frac{y+15}{-7} = \frac{z-9}{5}, \qquad (2.91)$$

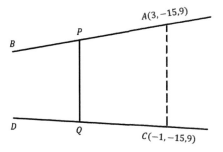

and

$$\frac{x+1}{2} = \frac{y-1}{1} = \frac{z-9}{-3}. \qquad (2.92)$$

Let $A\,(3, -15, 9)$, and $C\,(-1, 1, 9)$ be the points on the line (2.91) and (2.92)
Let l, m, n be the direction cosines of the shortest distance, then since the shortest distance is perpendicular to both the lines (2.91) and (2.92), we get

$$2l - 7m + 5n = 0 \quad 2l + m - 3n = 0$$

$$\therefore \frac{l}{21-5} = \frac{m}{+10+6} = \frac{n}{16}$$

$$\therefore \frac{l}{16} = \frac{m}{16} = \frac{n}{16}$$

$$\therefore l = m = n.$$

$\therefore$ The direction ratios of the shortest distance are $1, 1, 1$.

∴ The direction cosines of PQ are $\frac{1}{\sqrt{3}}, \frac{1}{\sqrt{3}}, \frac{1}{\sqrt{3}}$.

∴ The length of shortest distance = Projection of AC on the line PQ of shortest distance = $\frac{1}{\sqrt{3}}(-1-3) + \frac{1}{\sqrt{3}}(1+15) + \frac{1}{\sqrt{13}}(9-9) = \frac{12}{\sqrt{3}} = 4\sqrt{3}$.

The line of shortest distance is the line of intersection of the planes through (2.91) and the shortest distance and the plane through (2.92) and the shortest distance.

∴ Equation of the plane through (2.91) and the shortest distance is

$$\begin{vmatrix} x-3 & y+15 & z-9 \\ 2 & -7 & 5 \\ 1 & 1 & 1 \end{vmatrix} = 0.$$

∴ $(x-3)(-7-5) - (y+15)(2-5) + (z-9)(2+7) = 0$

∴ $-12x + 36 + 3y + 45 + 9z - 81 = 0$

∴ $-12x + 3y - 9z = 0$

∴ $4x - y - 3z = 0,$ \hfill (2.93)

and the equation of the plane through (2.92) and the shortest distance is

$$\begin{vmatrix} x+1 & y-1 & z-9 \\ 2 & 1 & -3 \\ 1 & 1 & 1 \end{vmatrix} = 0$$

∴ $(x+1)(1+3) - (y-1)(2+3) + (z-9)(2-1) = 0$

∴ $4x + 4 - 5y + 5 + z - 9 = 0$

∴ $4x - 5y + z = 0.$ \hfill (2.94)

Thus from (2.93) and (2.94), the equations of the shortest distance are

$$4x - y - 3z = 0, \ 4x - 5y + z = 0.$$

27) Find the length and equations of the line of the shortest distance between the lines $\frac{x+3}{-4} = \frac{y-6}{3} = \frac{z}{2}; \frac{x+2}{-4} = \frac{y}{1} = \frac{z-7}{1}$.

Sol. The given lines are

$$\frac{x+3}{-4} = \frac{y-6}{3} = \frac{z}{2} \hfill (2.95)$$

and

$$\frac{x+2}{-4} = \frac{y}{1} = \frac{z-7}{1}. \hfill (2.96)$$

2.10 Equation of Two Skew Lines in Symmetric Form

Let $A(-3, 6, 0)$ and $C(-2, 0, 7)$ be the points on the line (2.95) and (2.96).
Let l, m, n be the direction cosines of the shortest distance; then since the shortest distance is perpendicular to both the lines (2.95) and (2.96), we get

$$-4l + 3m + 2n = 0$$
$$-4l + m + n = 0$$
$$\therefore \frac{l}{3-2} = \frac{m}{-8+4} = \frac{n}{-4+2}$$
$$\therefore \frac{l}{1} = \frac{m}{-4} = \frac{n}{8}.$$

$\therefore$ The direction ratios of the shortest distance are $1, -4, 8$.
  $\therefore$ The direction cosine's of PQ are $\left(\frac{1}{9}, \frac{-4}{9}, \frac{8}{9}\right)$.
  $\therefore$ The length of shortest distance $= \frac{1}{9}(-2+3) - \frac{4}{9}(0-6) + \frac{8}{9}(7-0)$

$$= \frac{1}{9} + \frac{24}{9} + \frac{56}{9} = \frac{81}{9} = 9.$$

The line of shortest distance is the line of intersection of the planes through (2.95) and shortest distance and the plane through (2.96) and the shortest distance.
  $\therefore$ Equation of the plane through (2.95) and the shortest distance is

$$\begin{vmatrix} x-3 & y-6 & z \\ -4 & 3 & 2 \\ 1 & -4 & 8 \end{vmatrix} = 0.$$

$\therefore (x+3)(24+8) - (y-6)(-32-2) + z(16-3) = 0$
$\therefore 32x + 34y + 13z - 108 = 0 \quad (2.97)$

and the equation of the plane through (2.96) and the shortest distance is

$$\begin{vmatrix} x+2 & y & z-7 \\ -4 & 1 & 1 \\ 1 & -4 & 8 \end{vmatrix} = 0.$$

$(x+2)(8+4) - y(-32-1) + (z-7)(16-1) = 0$
$\therefore 12x + 33y + 15z - 81 = 0. \quad (2.98)$

Equations (2.97) and (2.98) are required equations of the line of the shortest distance.

28) Find the magnitude and equation of the line of the shortest distance between the lines $\frac{x}{4} = \frac{y+1}{3} = \frac{z-2}{2}$ and $5x - 2y - 3z + 60 = x - 3y + 2z - 3$.

Sol. The given equation of lines are

$$\frac{x}{4} = \frac{y+1}{3} = \frac{z-2}{2} \qquad (2.99)$$

and

$$\left.\begin{array}{l} 5x - 2y - 3z + 6 = 0 \\ x - 3y + 2z - 3 = 0 \end{array}\right\} . \qquad (2.100)$$

The equation of a plane through the line (2.100) is given by

$$5x - 2y - 3z + 6 + k(x - 3y + 2z - 3) = 0$$

$$(5+k)x + (-2-3k)y + (-3+2k)z + 6 - 3k = 0. \qquad (2.101)$$

If Equation (2.101) is parallel to the (2.99), then

$$4(5+k) + 3(-2-3k) + 2(-3+2k) = 0$$

$$20 + 4k - 6 - 9k - 6 + 4k = 0$$

$$-k + 8 = 0$$

$$\therefore k = 8.$$

$\therefore$ Equation (2.101) becomes,

$$13x - 26y + 13z - 18 = 0. \qquad (2.102)$$

Now, the shortest distance is the distance of the point $(0, -1, 2)$ from the plane (2.102) $= \frac{13(0) - 26(-1) + 13(2) - 18}{\sqrt{(13)^2 + (-26)^2 + (13)^2}} = \frac{34}{13\sqrt{6}} = \frac{17\sqrt{6}}{39}$.

$\therefore$ Equation of the plane through (2.99) and perpendicular to (2.102) is given by

$$\begin{vmatrix} x & y+1 & z-2 \\ 4 & 3 & 2 \\ 13 & -26 & 1 \end{vmatrix} = 0$$

$$\Rightarrow 13 \begin{vmatrix} x & y+1 & z-2 \\ 4 & 3 & 2 \\ 1 & -2 & 1 \end{vmatrix} = 0$$

$$\therefore 7x - 2y - 11z + 20 = 0. \qquad (2.103)$$

2.10 Equation of Two Skew Lines in Symmetric Form

If the Equation (2.101) is perpendicular to (2.102) then we get

$$13(5+k) + 26(2+k) + 13(-3+2k) = 0,$$

$$\therefore k = \frac{-2}{3}.$$

$\therefore$ Equation (2.101) becomes,

$$13x - 13z + 24 = 0. \tag{2.104}$$

Planes (2.103) and (2.104) give the equation of shortest distance.

28) Find the shortest distance between the $axis$ of z and the line $ax + by + cz + d = 0$ and $a^1x + b^1y + c^1z + d^1 = 0$.

Sol. The given equations of the line are

$$ax + by + cz + d = 0 \quad \text{and} \quad a^1x + b^1y + c^1z + d^1 = 0. \tag{2.105}$$

Let the equation of a plane through the given lines be

$$ax + by + cz + d + k\left(a^1x + b^1y + c^1z + d^1\right) = 0$$

i.e.,

$$\left(a + ka^1\right)x + \left(b + kb^1\right)y + \left(c + kc^1\right)z + d + kd^1 = 0. \tag{2.106}$$

If this line is parallel to z-axis then

$$\left(a + ka^1\right)(0) + \left(b + kb^1\right)(0) + \left(c + kc^1\right)(1) = 0$$

$$\therefore k = \frac{-c}{c^1}.$$

Substituting the value of k in Equation (2.106), we get the equation of the plane passing through a given line and parallel to z-axis as $\left(ac^1 - ca^1\right)x + \left(bc^1 - cb^1\right)y + dc^1 - cd^1 = 0$.

The shortest distance = the distance of this plane from the point $(0, 0, 1)$

$$= \frac{dc^1 - cd^1}{\sqrt{\left(ac^1 - ca^1\right)^2 + \left(bc^1 - b^1c\right)^2}}.$$

2.11 Intersection of Three Planes

Let the three planes be given by

$$a_1x + b_1y + c_1z + d_1 = 0, \qquad (2.107)$$

$$a_2x + b_2y + c_2z + d_2 = 0, \qquad (2.108)$$

$$a_3x + b_3y + c_3z + d_3 = 0. \qquad (2.109)$$

There are three cases in respect to the intersection of planes:

1) The three planes intersect in a common point (Fig (i))

2) The three planes have a line in a common (Fig (ii))

3) The three planes form a triangular prism (Fig (iii))

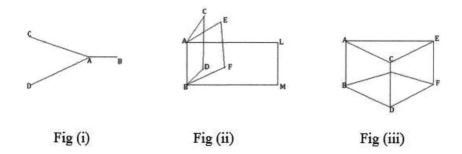

Fig (i) Fig (ii) Fig (iii)

Three planes are said to form a triangular prism if the three lines of intersection of the three planes, taken in pairs, are distinct and parallel.

To find the line of intersection of (2.107) and (2.108) in the symmetrical form

$$a_1x + b_1y + c_1z = 0;\ a_2x + b_2y + c_2z = 0$$

$$\therefore \frac{x}{b_1c_2 - b_2c_1} = \frac{y}{c_1a_2 - a_1c_2} = \frac{z}{a_1b_2 - a_2b_1}$$

$\therefore$ The direction ratios of the line of intersection of (2.107) and (2.108) are $b_1c_2 - b_2c_1, c_1a_2 - a_1c_2, a_1b_2 - a_2b_1$.

Substituting $z = 0$ in (2.107) and (2.108), we get

$$a_1x + b_1y + d_1 = 0$$

$$a_2x + b_2y + d_2 = 0$$

$$\therefore \frac{x}{b_1 d_2 - b_2 d_1} = \frac{y}{a_2 d_1 - a_1 d_2} = \frac{1}{a_1 b_2 - a_2 b_1}$$

$$\therefore x = \frac{b_1 d_2 - b_2 d_1}{a_1 b_2 - a_2 b_1}, y = \frac{a_2 d_1 - a_1 d_2}{a_1 b_2 - a_2 b_1}, z = 0.$$

$\therefore$ One point on the line of intersection of (2.107) and (2.108) is

$$\left(\frac{b_1 d_2 - b_2 d_1}{a_1 b_2 - a_2 b_1}, \frac{a_2 d_1 - a_1 d_2}{a_1 b_2 - a_2 b_1}, 0 \right).$$

$\therefore$ The equations of the line of intersection of (2.107) and (2.108) in the symmetrical form are

$$\frac{x - \frac{b_1 d_2 - b_2 d_1}{a_1 b_2 - b_1 a_2}}{b_1 c_2 - b_2 c_1} = \frac{y - \frac{a_2 d_1 - a_1 d_2}{a_1 b_2 - a_2 b_1}}{c_1 a_2 - a_1 c_2} = \frac{z - 0}{a_1 b_2 - a_2 b_1} \quad (2.110)$$

(i) Condition for the three planes to meet at a point:

The three planes (2.107), (2.108), and (2.109) will intersect in a point if the line of intersection of (2.107) and (2.108).

i.e., the line (2.110) cuts the plane (2.109) at a point, and for this (2.110) should not be parallel to the plane (2.109).

i.e., $al + bm + cn \neq 0$

i.e., $a_3 (b_1 c_2 - b_2 c_1) + b_3 (a_2 c_1 - a_1 c_2) + c_3 (a_1 b_2 - a_2 b_1) \neq 0$

$$\Rightarrow \Delta_4 = \begin{vmatrix} a_1 & b_1 & c_1 \\ a_2 & b_2 & c_2 \\ a_3 & b_3 & c_3 \end{vmatrix} \neq 0.$$

Solving the Equations (2.107), (2.108), and (2.109) by determinants, we get

$$\frac{x}{\begin{vmatrix} b_1 & c_1 & d_1 \\ b_2 & c_2 & d_2 \\ b_3 & c_3 & d_3 \end{vmatrix}} = \frac{y}{\begin{vmatrix} a_1 & c_1 & d_1 \\ a_2 & c_2 & d_2 \\ a_3 & c_3 & d_3 \end{vmatrix}} = \frac{z}{\begin{vmatrix} a_1 & b_1 & d_1 \\ a_2 & b_2 & d_2 \\ a_3 & b_3 & d_3 \end{vmatrix}} = \frac{-1}{\begin{vmatrix} a_1 & b_1 & c_1 \\ a_2 & b_2 & c_2 \\ a_3 & b_3 & c_3 \end{vmatrix}}$$

$$\therefore \frac{x}{\Delta_1} = \frac{-y}{\Delta_2} = \frac{z}{\Delta_3} = \frac{-1}{\Delta_4}$$

$$\therefore x = \frac{-\Delta_1}{\Delta_4}, y = \frac{\Delta_2}{\Delta_2}, \text{ and } z = \frac{-\Delta_3}{\Delta_4}.$$

If the planes intersect at a point, the coordinates of their point of intersection are finite.

∴ The $\Delta_4 \neq 0$ which is the required condition.

(ii) Condition for the three planes to form a triangular plane:

If the planes (2.107), (2.108), and (2.109) form a triangular prism, then the line of intersection of (2.107) and (2.108)
i.e., the line (2.110) should be parallel to the plane (2.109);

(a) $al + bm + cn = 0$ and

(b) one point on the line (2.110) should not lie in the plane (2.109).

Now (a) $al + bm + cn = 0$ if

$$a_3 (b_1 c_2 - c_1 b_2) + b_3 (c_1 a_2 - a_1 c_2) + c_3 (a_1 b_2 - b_1 a_2) = 0$$

i.e., $\Delta = \begin{vmatrix} a_1 & b_1 & c_1 \\ a_2 & b_2 & c_2 \\ a_3 & b_3 & c_3 \end{vmatrix} = 0$

and (b) one point $\left(\frac{b_1 d_2 - d_1 b_2}{a_1 b_2 - a_2 b_1}, \frac{a_2 d_1 - d_2 a_1}{a_1 b_2 - b_1 a_2}, 0 \right)$ on the line (2.110) will not lie on the plane (2.109) if

$$a_3 \left(\frac{b_1 d_2 - d_1 b_2}{a_1 b_2 - a_2 b_1} \right) + b_3 \left(\frac{a_2 d_1 - d_2 a_1}{a_1 b_2 - b_1 a_2} \right) + c_3 (0) + d_3 \neq 0$$

i.e., $a_3 (b_1 d_2 - d_1 b_2) + b_3 (a_2 d_1 - d_2 a_1) + d_3 (a_1 b_2 - a_2 b_1) \neq 0$

i.e., if $\Delta_4 = \begin{vmatrix} a_1 & b_1 & d_1 \\ a_2 & b_2 & d_2 \\ a_3 & b_3 & d_3 \end{vmatrix} = 0.$

(ii) Let the point $\left(\frac{b_1 d_2 - d_1 b_2}{a_1 b_2 - a_2 b_1}, \frac{a_2 d_1 - d_2 a_1}{a_1 b_2 - b_1 a_2}, 0 \right)$ on the line (2.110) lies on the plane (2.109) if

$$a_3 \left(\frac{b_1 d_2 - d_1 b_2}{a_1 b_2 - a_2 b_1} \right) + b_3 \left(\frac{d_1 a_2 - a_1 d_2}{a_1 b_2 - a_2 b_1} \right) + c_1 (0) + d_1 = 0$$

∴ $a_3 (b_1 d_2 - d_1 b_2) + b_3 (d_1 a_2 - a_1 d_2) + d_1 (a_1 b_2 - a_2 b_1) = 0$

∴ $\Delta_3 = \begin{vmatrix} a_1 & b_1 & d_1 \\ a_2 & b_2 & d_2 \\ a_3 & b_3 & d_3 \end{vmatrix} = 0.$

Hence the required conditions are $\Delta_4 = 0$ and $\Delta_1 = \Delta_2 = \Delta_3 = 0$.

2.11 Intersection of Three Planes

The Working rule for finding the nature of the intersection of the three planes:

Let the three planes be

$$a_1x + b_1y + c_1z + d_1 = 0,$$
$$a_2x + b_2y + c_2z + d_2 = 0,$$
$$a_3x + b_3y + c_3z + d_3 = 0.$$

Step I: Write down the coefficients in the equations to get the rectangular array

$$\Delta = \begin{vmatrix} a_1 & b_1 & c_1 & d_1 \\ a_2 & b_2 & c_2 & d_2 \\ a_3 & b_3 & c_3 & d_3 \end{vmatrix}.$$

Step II: Omit the fourth column to get the determinant

$$\Delta_4 = \begin{vmatrix} a_1 & b_1 & c_1 \\ a_2 & b_2 & c_2 \\ a_3 & b_3 & c_3 \end{vmatrix}.$$

Now if $\Delta_4 \neq 0$, the planes intersect in a point.

Step III: But if $\Delta_4 = 0$, then omit the third column in Δ to get

$$\Delta_3 = \begin{vmatrix} a_1 & b_1 & d_1 \\ a_2 & b_2 & d_2 \\ a_3 & b_3 & d_3 \end{vmatrix}.$$

If $\Delta_3 \neq 0$, the planes form a triangular prism.
If $\Delta_3 = 0$, the planes intersect in a line.

30) Examine the nature of the intersection of the following sets of planes:

1) $2x + 3y - z - 2 = 0$, $3x + 3y + z - 4 = 0$, $x - y + 2z - 5 = 0$
2) $4x - 5y - 2z - 2 = 0$, $5x - 4y + 2z + 2 = 0$, $2x + 2y + 8z - 1 = 0$
3) $5x + 3y + 7z - 4 = 0$, $3x + 26y + 2z - 9 = 0$, $7x + 2y + 10z - 5 = 0$

Sol. 1) The given equations of planes are

$$2x + 3y - z - 2 = 0$$

84 Straight Line

$$3x + 3y + z - 4 = 0$$
$$x - y + 2z - 5 = 0$$

$$\therefore \Delta_4 = \begin{vmatrix} 2 & 3 & -1 \\ 3 & 3 & 1 \\ 1 & -1 & 2 \end{vmatrix} = 2(3+1) - 3(6-1) - 1(-3-3) = -1 \neq 0$$

∴ The given three planes intersect at a point.

2) The given equation of planes are

$$4x - 5y - 2z - 2 = 0$$
$$5x - 4y + 2z + 2 = 0$$
$$2x + 2y + 8z - 1 = 0$$

$$\therefore \Delta_4 = \begin{vmatrix} 4 & -5 & -2 \\ 5 & -4 & 2 \\ 2 & 2 & 8 \end{vmatrix} = 4(-32-4) + 5(40-4) - 2(10+8)$$

$$= -144 + 180 - 36 = 0$$

$$\Delta_3 = \begin{vmatrix} 4 & -5 & -2 \\ 5 & -4 & 2 \\ 2 & 2 & 8 \end{vmatrix} = 4(4-4) + 5(-5-4) - 2(10+8) \neq 0$$

Since $\Delta_4 = 0$ and $\Delta_3 \neq 0$
∴ The three planes form a triangular prism.

1) The given equations of the planes are

$$5x + 3y + 7z - 4 = 0$$
$$3x + 26y + 2z - 9 = 0$$
$$7x + 2y + 10z - 5 = 0$$

$$\therefore \Delta_4 = \begin{vmatrix} 5 & 3 & 7 \\ 3 & 26 & 2 \\ 7 & 2 & 10 \end{vmatrix} = 5(260-4) - 3(30-14) + 7(6-182) = 0$$

$$\Delta_3 = \begin{vmatrix} 5 & 3 & -4 \\ 3 & 26 & -9 \\ 7 & 2 & -5 \end{vmatrix} = 5(-130+18) - 3(-15+63) - 4(6-182)$$

$$= 0$$

Since $\Delta_4 = 0$ and $\Delta_3 = 0$
∴ The three planes intersect in a line.

31) Show that the planes $ax + hy + gz = 0$, $hx + by + fz = 0$, $gz + fy + cz = 0$ have a common line of the intersection if $\Delta = \begin{vmatrix} a & h & g \\ h & b & f \\ g & f & c \end{vmatrix} = 0$ and the direction ratios of the line satisfy the equations $\frac{l^2}{\frac{\partial \Delta}{\partial a}} = \frac{m^2}{\frac{\partial \Delta}{\partial b}} = \frac{n^2}{\frac{\partial \Delta}{\partial c}}$.

Sol. Let given equations of plane be

$$ax + hy + gz = 0, \tag{2.111}$$

$$hx + by + fz = 0, \tag{2.112}$$

$$gx + fy + cz = 0. \tag{2.113}$$

Let l, m, n be the direction cosines of the line of intersection of the planes (2.111) and (2.112) then $al + hm + gn = 0$ and $hl + bm + fn = 0$.

$$\Rightarrow \frac{l}{hf - bg} = \frac{m}{gl - af} = \frac{n}{ab - h^2}. \tag{2.114}$$

But the three planes pass through a common point $(0, 0, 0)$.
∴ The line of intersection of planes (2.111) and (2.112) will be in the plane (2.113) if $g(hf - bg) + f(gh - af) + c(ab - h^2) = 0$

$$\therefore abc + 2fgh - af^2 - bg^2 - ch^2 = 0$$

$$\Delta = \begin{vmatrix} a & h & g \\ h & b & f \\ g & f & c \end{vmatrix} = 0.$$

From equation (2.111), we get

$$\frac{l^2}{(hf - bg)^2} = \frac{m^2}{(gh - af)^2} = \frac{n^2}{(ab - h^2)^2} \tag{2.115}$$

$$(hf - bg)^2 = h^2 f^2 + b^2 g^2 - 2hfbg$$
$$= h^2 f^2 + b^2 g^2 - b(af^2 + bg^2 + ch^2 - abc)$$
$$= (ab - h^2)(bc - f^2)$$

(From (2.114))

86 Straight Line

Similarly, $(gh - af) = (ab - h^2)(ca - g^2)$
Substituting these values in (2.115), we get

$$\frac{l^2}{(ab - h^2)(bc - f^2)} = \frac{m^2}{(ab - h^2)(ca - g^2)} = \frac{n^2}{(ab - h^2)^2}$$

$$\therefore \frac{l^2}{bc - f^2} = \frac{m^2}{ca - g^2} = \frac{n^2}{ab - h^2}$$

$$\therefore \frac{l^2}{\frac{\partial \Delta}{\partial a}} = \frac{m^2}{\frac{\partial \Delta}{\partial b}} = \frac{n^2}{\frac{\partial \Delta}{\partial c}}$$

where $\Delta = abc + 2fgh - af^2 - bg^2 - ch^2$
$\therefore \frac{\partial \Delta}{\partial a} = bc - f^2$, $\frac{\partial \Delta}{\partial b} = ac - g^2$, and $\frac{\partial \Delta}{\partial c} = ab - h^2$.

32) Prove that the planes $x = cy + bz$, $y = az + cx$ and $z = bx + ay$ pass through one line if $a^2 + b^2 + c^2 + 2abc = 1$. Show that the equations of this line are $\frac{x}{\sqrt{1-a^2}} = \frac{y}{\sqrt{1-b^2}} = \frac{z}{\sqrt{1-c^2}}$.

Sol. The three given planes can be written as

$$x - cy - bz = 0, \qquad (2.116)$$

$$cx - y + az = 0, \qquad (2.117)$$

$$bx + ay - z = 0. \qquad (2.118)$$

Solving (2.116) and (2.117) we get a line of intersection of the plane (2.116) and (2.117),

$$\frac{x}{-ac - b} = \frac{y}{-bc - a} = \frac{z}{-1 + c^2}$$

i.e.,

$$\frac{x}{ac + b} = \frac{y}{bc + a} = \frac{z}{1 - c^2} \qquad (2.119)$$

The three planes will intersect in a line if the line of intersection of (2.116) and (2.117).
i.e., the line (2.119) lies in the plane (2.118).
The point $(0, 0, 0)$ on (2.119) satisfies (2.118).
$\therefore$ The line (2.119) will lie in the plane (2.118) if $al + bm + cn = 0$
i.e., if $b(ac + b) + a(bc + a) - 1(1 - c^2) = 0$

$$\therefore a^2 + b^2 + c^2 + 2abc = 1 \qquad (2.120)$$

which is the required condition.

2.11 Intersection of Three Planes

Now to find the line of intersection (2.119) in the given form:

$$ac + b = \sqrt{(ac+b)^2} = \sqrt{a^2c^2 + b^2 + 2abc} = \sqrt{a^2c^2 + 1 - a^2 - c^2}$$
(From (2.120))

$$\therefore ac + b = \sqrt{(1-a^2)(1-c^2)}$$

Similarly, $bc + a = \sqrt{(1-b^2)(1-c^2)}$

Substituting these values in (2.119), the required line of intersection is

$$\frac{x}{\sqrt{(1-a^2)(1-c^2)}} = \frac{y}{\sqrt{(1-b^2)(1-c^2)}} = \frac{z}{1-c^2}$$

$$\therefore \frac{x}{\sqrt{1-a^2}} = \frac{y}{\sqrt{1-b^2}} = \frac{z}{\sqrt{1-c^2}}.$$

Exercise:

1) Find the equations of the straight line joining the points $(-2, 1, 3)$ and $(3, 1, -2)$.

   Answer: $\frac{x+2}{1} = \frac{y-1}{0} = \frac{z-3}{-1}$

2) Find the equations of a straight line through the point $(3, 1, -6)$ and parallel to each of the planes $x + y + 2z - 4 = 0$ and $2x - 3y + z + 5 = 0$.

   Answer: $\frac{x-3}{7} = \frac{y-1}{3} = \frac{z+6}{-5}$

3) Find the equations of the plane through $(3, 1, -1)$ perpendicular to the line of the planes $3x + 4y + 7z + 4 = 0$ and $x - y + 2z - 3 = 0$.

   Answer: $15x + y - 7z - 53 = 0$

4) Find the equations of the line through the point $(1, 1, 1)$ and perpendicular to the line $x - 2y + z = 2$, $4x + 3y - z + 1 = 0$.

   Answer: $x - 5y - 11z + 15 = 0$

5) Find the symmetrical form of the line $3x + 2y - z - 5 = 0 = x + y - 2z - 3$.

   Answer: $\frac{x+1}{-5} = \frac{y-4}{7} = \frac{z}{1}$

6) Find the angle between the lines $\frac{x}{1} = \frac{y}{0} = \frac{z}{-1}$ and $\frac{x}{3} = \frac{y}{4} = \frac{z}{5}$.

   Answer: $\theta = \cos^{-1}\left(\frac{1}{5}\right)$

7) Find the angle between the lines in which the planes $3x - 7y - 5z = 1$, $5x - 13y + 3z + 2 = 0$ cut the plane $8x - 11y + z = 0$.

   Answer: $90°$

8) Find the point where the line $\frac{x-1}{2} = \frac{y-2}{-3} = \frac{z+3}{4}$ meets the plane $2x + 4y - z + 1 = 0$.

Answer: $\left(\frac{10}{3}, \frac{-3}{2}, \frac{5}{3}\right)$

9) Show that the equation of the plane which passes through the line $\frac{x-1}{3} = \frac{y+6}{4} = \frac{z+1}{2}$ and is parallel to the line $\frac{x-2}{2} = \frac{y-1}{-3} = \frac{z+4}{5}$, is $26x - 11y - 17x - 109 = 0$.

10) Find the distance of the point $(-1, -5, -10)$ from the point of intersection of the line $\frac{x-2}{3} = \frac{y+1}{4} = \frac{z-2}{12}$ and the plane $x - y + z = 5$.

Answer: 13

11) Show that the lines $\frac{x+3}{2} = \frac{y+5}{3} = \frac{z-7}{-3}, \frac{x+1}{4} = \frac{y+1}{5} = \frac{z+1}{-1}$ are coplanar and find the equation of the plane containing them.

Answer: $6x - 5y - z = 0$

12) Prove that the lines $\frac{x-1}{2} = \frac{y+1}{-3} = \frac{z+10}{8}$; ; $\frac{x-4}{1} = \frac{y+3}{-4} = \frac{z+1}{7}$ intersect. Find also their point of intersection and the plane through them.

Answer: $(5, -7, 6)$; ; $11x - 6y - 5z = 67$

13) Show that the plane containing the two parallel lines $x - 4 = \frac{y-3}{-4} = \frac{z-2}{5}$; ; $x - 3 = \frac{y+2}{-4} = \frac{z}{5}$ is $11x - y - 3z = 35$.

14) Prove that the line $\frac{x-3}{2} = \frac{y-4}{3} = \frac{z-5}{4}$ lies in the plane $4x + 4y - 5z - 3 = 0$.

15) Find the equation of the plane which passes through the point $(1, 2, -1)$ and which contains the line $\frac{x+1}{2} = \frac{y-1}{3} = \frac{z+2}{-1}$.

Answer: $x - y - z = 0$

16) Find the length and the equations of the shortest distance between $5x - y - z = 0 = x - 2y + z + 3$; ; $7x - 4y - 2z = 0 = x - y + z - 3$.

Answer: $\frac{13}{\sqrt{75}}$; ; $17x + 20y - 19z - 39 = 0 = 8x + 5y - 31z + 67$

17) Examine the nature of the intersection of the following sets of planes:

(i) $4x + 3y + 2z + 7 = 0$, $2x + y - 4z + 1 = 0$, $x - 7z - 7 = 0$
(ii) $x + y + z + 3 = 0$, $3x + y - 2z + 2 = 0$, $2x + 4y + 7z - 7 = 0$

Answer: (i) Line (ii) Prism

3

Sphere

3.1 Definition

The sphere is a locus of a point that moves so that its distance from a fixed point always remains constant.

The fixed point is called the center of the sphere and the constant distance is called the radius of the sphere.

3.2 Equation of Sphere in Vector Form

Consider a sphere S with center C and radius 'a'. Let P be any point on the sphere. Let $\vec{c}$ and $\vec{r}$ be the position vectors of C and P respectively with respect to O; the origin of reference.

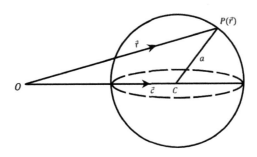

$$\therefore \overrightarrow{CP} = \overrightarrow{OP} - \overrightarrow{OC}$$
$$\therefore \overrightarrow{CP} = \vec{r} - \vec{c}$$

But radius $CP = \left|\overrightarrow{CP}\right| = a.$

$$\therefore |\vec{r} - \vec{c}| = a \qquad (3.1)$$

$$\therefore |\vec{r} - \vec{c}|^2 = a^2$$
$$\therefore (\vec{r} - \vec{c}) \cdot (\vec{r} - \vec{c}) = a^2$$
$$\therefore \vec{r} \cdot \vec{r} - \vec{r} \cdot \vec{c} - \vec{c} \cdot \vec{r} + \vec{c} \cdot \vec{c} = a^2$$
$$\therefore |\vec{r}|^2 - 2\vec{r} \cdot \vec{c} + |\vec{c}|^2 = a^2$$
$$\therefore |\vec{r}|^2 - 2\vec{r} \cdot \vec{c} + d = 0$$
$$\therefore |\vec{r}|^2 - 2\vec{r} \cdot \vec{c} + |\vec{c}|^2 = a^2 \tag{3.2}$$

where $d = |\vec{c}|^2 - a^2$
is the equation of sphere in vector form.

Case 1): If O lies on the sphere, then $OC = OP$;
i.e., $c = a \Rightarrow d = 0$
$\therefore$ The Equation of sphere is $|\vec{r}|^2 - 2\vec{r} \cdot \vec{c} = 0$

Case 2): If the center C is at the origin O, then $c = 0$ and $d = -a^2$.
$\therefore$ Equation of sphere is $|\vec{r}|^2 - a^2 = 0$.

$$|\vec{r}|^2 = a^2$$
$$|\vec{r}| = a.$$

Equation of sphere in the cartesian form:

Consider a sphere S with center $C(\alpha, \beta, \gamma)$, and radius 'a'. Let $P(x, y, z)$ be any point on the sphere. Let $\vec{r}$ and $\vec{c}$ be the position vector of P and C respectively with respect to O; the origin of reference.

$$\therefore \vec{r} = x\vec{i} + y\vec{j} + z\vec{k}$$
$$\vec{c} = \alpha\vec{i} + \beta\vec{j} + \gamma\vec{k}.$$

The equation in vector form is $|\vec{r} - \vec{c}| = a$.

$$\therefore \left|(x-\alpha)\vec{i} + (y-\beta)\vec{j} + (z-\gamma)\vec{k}\right| = a$$
$$\therefore (x-\alpha)^2 + (y-\beta)^2 + (z-\gamma)^2 = a^2$$

which is the equation of sphere S in cartesian form with center $C(\alpha, \beta, \gamma)$, and radius 'a'.

Remark:

When center C is origin i.e., $\alpha = \beta = \gamma = 0$ then the equation of the sphere is $x^2 + y^2 + z^2 = a^2$.

3.3 General Equation of the Sphere

The general equation of the sphere is

$$x^2 + y^2 + z^2 + 2ux + 2vy + 2wz + d = 0 \qquad (3.3)$$

$$\therefore x^2 + 2ux + u^2 + y^2 + 2vy + v^2 + z^2 + 2wz + w^2 + d = u^2 + v^2 + w^2$$

$$\therefore (x+u)^2 + (y+v)^2 + (z+w)^2 = u^2 + v^2 + w^2 - d$$

$$\therefore (x+u)^2 + (y+v)^2 + (z+w)^2 = a^2$$

where $a^2 = u^2 + v^2 + w^2 - d$
which is the equation of sphere with the center $(-u, -v, -w)$ and radius

$$a = \sqrt{u^2 + v^2 + w^2 - d}.$$

Remark:

1) Characteristics of the equation of the sphere:

   1.1 The equation of sphere is a second-degree equation in $x, y,$ and z.

   1.2 Coefficients of $x^2, y^2,$ and z^2 are the same.

   1.3 The products terms $xy, yz,$ and zx are absent.

2) Equation of the sphere with the center origin and radius a is

$$x^2 + y^2 + z^2 = a^2.$$

3) Make the coefficients of $x^2, y^2,$ and z^2 one if they are not before finding the center and radius of the sphere.

3.4 Equation of Sphere Whose End-Points of a Diameter are Given

Let O be the origin of the vector. Let $\overrightarrow{AB}$ be a diameter of a sphere S with center C. Let $\vec{a}$ and $\vec{b}$ be position vectors of points A and B respectively. Let P be any point on the sphere with position vector $\vec{r}$. $\angle APB$ is an angle inscribed in a semi-circle.

92 Sphere

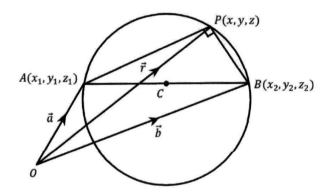

∴ ∠APB is the right angle.
∴ The vectors $\overrightarrow{AP}$ and $\overrightarrow{BP}$ are perpendicular.

$$\therefore \overrightarrow{AP} = \overrightarrow{OP} - \overrightarrow{OA} = \vec{r} - \vec{a}$$
$$\overrightarrow{BP} = \overrightarrow{OP} - \overrightarrow{OB} = \vec{r} - \vec{b}$$

∴ By condition of perpendicularity;

$$\overrightarrow{AP}.\overrightarrow{BP} = 0$$
$$\therefore (\vec{r} - \vec{a}).(\vec{r} - \vec{b}) = 0 \qquad (3.4)$$

which is the equation of a sphere whose end-points of a diameter are given.

Let $P(x, y, z)$; $A(x_1, y_1, z_1)$; $B(x_2, y_2, z_2)$ be coordinates of points then

$$\vec{r} = x\vec{i} + y\vec{j} + z\vec{k}$$
$$\vec{a} = x_1\vec{i} + y_1\vec{j} + z_1\vec{k}$$
$$\vec{b} = x_2\vec{i} + y_2\vec{j} + z_2\vec{k}.$$

∴ Equation (3.4) becomes;

$$\left[(x - x_1)\vec{i} + (y - y_1)\vec{j} + (z - z_1)\vec{k}\right].$$
$$\left[(x - x_2)\vec{i} + (y - y_2)\vec{j} + (z - z_2)\vec{k}\right] = 0$$
$$\therefore (x - x_1)(x - x_2) + (y - y_1)(y - y_2) + (z - z_1)(z - z_2) = 0$$

is the Cartesian equation of sphere with diameter AB.

3.5 Equation of a Sphere Passing through the Four Points

Let $(x_1, y_1, z_1), (x_2, y_2, z_2), (x_3, y_3, z_3)$ and (x_4, y_4, z_4) be the four non-coplanar points. Let the required equation of a sphere be

$$x^2 + y^2 + z^2 + 2ux + 2vy + 2wz + d = 0. \qquad (3.5)$$

It passes through $(x_1, y_1, z_1), (x_2, y_2, z_2), (x_3, y_3, z_3)$ and (x_4, y_4, z_4) so we get

$$x_1^2 + y_1^2 + z_1^2 + 2ux_1 + 2vy_1 + 2wz_1 + d = 0, \qquad (3.6)$$
$$x_2^2 + y_2^2 + z_2^2 + 2ux_2 + 2vy_2 + 2wz_2 + d = 0, \qquad (3.7)$$
$$x_3^2 + y_3^2 + z_3^2 + 2ux_3 + 2vy_3 + 2wz_3 + d = 0, \qquad (3.8)$$
$$x_4^2 + y_4^2 + z_4^2 + 2ux_4 + 2vy_4 + 2wz_4 + d = 0. \qquad (3.9)$$

By solving Equations (3.6), (3.7), (3.8), and (3.9) we get the value of u, v, w, and d and we substitute u, v, w and d in equation (3.5) which is the required equation of the sphere passing through four points.

1) Find the equation of the sphere whose center is $(-1, 2, 3)$, and radius 3.

Sol. Let $P(x, y, z)$ be any point on the sphere whose center is $C(-1, 2, 3)$ and radius r is 3.

$$\therefore CP = 3$$
$$\therefore CP^2 = 9$$
$$\therefore (x+1)^2 + (y-2)^2 + (z-3)^2 = 9$$
$$x^2 + y^2 + z^2 + 2x - 4y - 6z + 5 = 0.$$

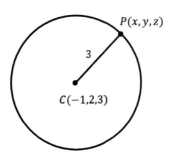

2) Find the equation of the sphere whose center is $(\frac{1}{2}, -1, \frac{-1}{2})$ and radius 1.

Sol. Let $P(x, y, z)$ be any point on the sphere whose center is $(\frac{1}{2}, -1, \frac{-1}{2})$ and radius r is 1.

$$\therefore CP = 1$$
$$\therefore CP^2 = 1$$
$$\left(x - \tfrac{1}{2}\right)^2 + (y+1)^2 + \left(z + \tfrac{1}{2}\right)^2 = 1$$
$$x^2 - x + \tfrac{1}{4} + y^2 + 2y + 1 + z^2 + z + \tfrac{1}{4} = 1$$
$$x^2 + y^2 + z^2 - x + 2y + z + \tfrac{1}{2} = 0$$
$$\therefore 2x^2 + 2y^2 + 2z^2 - 2x + 4y + 2z + 1 = 0.$$

3) Find the center and radius of the following spheres:
(i) $x^2 + y^2 + z^2 + 2x - 4y - 6z + 5 = 0$
Comparing with general equations of the sphere

$$x^2 + y^2 + z^2 + 2ux + 2vy + 2wz + d = 0$$
$$\therefore 2u = 2; \, 2v = -4; \, 2w = -6; \, d = 5$$
$$\therefore u = 1; \, v = -2; \, w = -3; \, d = 5$$

$\therefore$ Center C is $(-u, -v, -w) = (-1, 2, 3)$
Radius $r = \sqrt{u^2 + v^2 + w^2 - d} = \sqrt{(1)^2 + (-2)^2 + (-3)^2 - 5} = 3.$

(ii) $2x^2 + 2y^2 + 2z^2 + 4x - 2y - 2z = 5$
Comparing with general equations of the sphere

$$x^2 + y^2 + z^2 + 2ux + 2vy + 2wz + d = 0$$
$$x^2 + y^2 + z^2 + 2x - y - z - \tfrac{5}{2} = 0$$
$$\therefore 2u = 2; \quad 2v = -1; \quad 2w = -1; \quad d = \tfrac{-5}{2}$$
$$\therefore u = 1; \quad v = \tfrac{-1}{2}; \quad w = \tfrac{-1}{2}; \quad d = \tfrac{-5}{2}$$

$\therefore$ Center C is $(-u, -v, -w) = \left(-1, \tfrac{1}{2}, \tfrac{1}{2}\right).$

$$r = \sqrt{u^2 + v^2 + w^2 - d} = \sqrt{1 + \tfrac{1}{4} + \tfrac{1}{4} + \tfrac{5}{2}} = 2.$$

4) Obtain the equation of the sphere whose endpoints of the diameter are $A\,(2, -3, 4)$ and $B\,(-5, 6, -7)$.

Sol. Suppose $P(x, y, z)$ be any point on the sphere.

3.5 Equation of a Sphere Passing through the Four Points

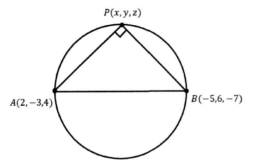

PA and PB are perpendiculars.

$\therefore$ By condition of perpendicularity; the equation of a sphere whose diameter end-points are given is

$$(x-2)(x+5) + (y+3)(y-6) + (z-4)(z+7) = 0$$
$$\therefore x^2 + 3x - 10 + y^2 - 3y - 18 + z^2 + 3z - 28 = 0$$
$$\therefore x^2 + y^2 + z^2 + 3x - 3y + 3z - 56 = 0.$$

5) Find the equation of the sphere through the points (0,0,0); (1,0,0); (0,2,0) and (0,0,3) and also find center.

Sol. Suppose the sphere and radius of the sphere passing through the given points is

$$x^2 + y^2 + z^2 + 2ux + 2vy + 2wz + d = 0. \qquad (3.10)$$

It passes through (0, 0, 0)

$$\therefore 0 + 0 + 0 + 0 + 0 + 0 + d = 0$$
$$\therefore d = 0.$$

It passes through (1, 0, 0)

$$\therefore 1 + 0 + 0 + 2u + 0 + 0 + d = 0$$
$$\therefore 2u = 1 \Rightarrow u = \tfrac{-1}{2}.$$

It passes through (0, 2, 0)

$$\therefore 0 + 4 + 0 + 0 + 4v + 0 + d = 0$$
$$\therefore 4v + 4 = 0 \Rightarrow v = -1.$$

It passes through (0, 0, 3)

$$\therefore 0+0+9+0+0+6w+d=0$$
$$\therefore 6w=-9 \Rightarrow w=\tfrac{-3}{2}.$$

Substituting the values of u, v, w, and d in (3.10), we get

$$x^2+y^2+z^2-x-2y-3z=0$$

Center $C(-u,-v,-w) = \left(\tfrac{1}{2}, 1, \tfrac{3}{2}\right)$

and radius $r = \sqrt{u^2+v^2+w^2-d} = \sqrt{\tfrac{1}{4}+1+\tfrac{9}{4}} = \sqrt{\tfrac{14}{4}} = \sqrt{\tfrac{7}{2}}$.

6) Find the equation of the sphere through the points $(0,0,0)$; $(-a,b,c)$; $(a,-b,c)$; $(a,b,-c)$.

Sol. Suppose the sphere passing through the given points is

$$x^2+y^2+z^2+2ux+2vy+2wz+d=0. \qquad (3.11)$$

It passes through (0, 0, 0)

$$\therefore 0+0+0+0+0+0+d=0$$
$$\therefore d=0.$$

It passes through $(-a, b, c)$

$$\therefore a^2+b^2+c^2-2ua+2vb+2wc=0$$
$$\therefore -2ua+2vb+2wc=-\left(a^2+b^2+c^2\right). \qquad (3.12)$$

It passes through $(a, -b, c)$

$$\therefore a^2+b^2+c^2+2au-2bv+2cw=0$$
$$\therefore 2au-2bv+2cw=-\left(a^2+b^2+c^2\right). \qquad (3.13)$$

It passes through $(a, b, -c)$

$$\therefore a^2+b^2+c^2+2au+2bv-2cw=0$$
$$\therefore 2au+2bv-2cw=-\left(a^2+b^2+c^2\right) \qquad (3.14)$$

Solving (3.12) and (3.13), we get

$$-2au+2bv+2cw=-\left(a^2+b^2+c^2\right)$$

3.5 Equation of a Sphere Passing through the Four Points

$$2au - 2bv + 2cw = -(a^2 + b^2 + c^2)$$
$$\therefore 4cw = -2(a^2 + b^2 + c^2)$$
$$\therefore w = -\frac{(a^2 + b^2 + c^2)}{2c}.$$

Solving (3.13) and (3.14), we get

$$2au - 2bv + 2cw = -(a^2 + b^2 + c^2)$$
$$2au + 2bv - 2cw = -(a^2+)$$
$$\therefore 4au = -2(a^2 + b^2 + c^2)$$

Solving (3.12) and (3.14), we get

$$-2au + 2bv + 2cw = -(a^2 + b^2 + c^2)$$
$$2au + 2bv - 2cw = -(a^2 + b^2 + c^2)$$
$$\therefore 4bv = -2(a^2 + b^2 + c^2)$$
$$\therefore v = -\frac{(a^2 + b^2 + c^2)}{2b}.$$

$\therefore$ Equation (3.11) becomes,

$$x^2 + y^2 + z^2 + 2x\left(\frac{-(a^2+b^2+c^2)}{2a}\right) + 2y\left(\frac{-(a^2+b^2+c^2)}{2b}\right)$$
$$+ 2z\left(\frac{-(a^2+b^2+c^2)}{2c}\right) = 0$$
$$\therefore (x^2 + y^2 + z^2) - \frac{x}{a}(a^2 + b^2 + c^2) - \frac{y}{b}(a^2 + b^2 + c^2)$$
$$- \frac{z}{c}(a^2 + b^2 + c^2) = 0$$
$$\therefore \frac{x^2+y^2+z^2}{a^2+b^2+c^2} - \left(\frac{x}{a} + \frac{y}{b} + \frac{z}{c}\right) = 0.$$

7) Obtain the equation of the sphere circumscribing the tetrahedron whose faces are $x = 0, y = 0, z = 0; \frac{x}{a} + \frac{y}{b} + \frac{z}{c} = 1$.

Sol. Four faces of the tetrahedron are

$$x = 0; y = 0; z = 0 \text{ and } \frac{x}{a} + \frac{y}{b} + \frac{z}{c} = 1$$

Solving three equations at a time we get points $O(0,0,0)$; $A(a,0,0)$; $B(0,b,0)$ and $C(0,0,c)$ which are four vertices of a tetrahedron.

Sphere passes through all four points.
The general equation of sphere is

$$\therefore x^2 + y^2 + z^2 + 2ux + 2vy + 2wz + d = 0. \qquad (3.15)$$

It passes through O $(0, 0, 0)$

$$\therefore 0 + 0 + 0 + 0 + 0 + 0 + d = 0$$
$$\therefore d = 0.$$

It passes through $A\,(a, 0,\ 0)$

$$\therefore a^2 + 0 + 0 + 2au + 0 + 0 + 0 = 0$$
$$\therefore a^2 + 2au = 0$$
$$\therefore u = \frac{-a}{2}.$$

It passes through $B\,(0, b, 0)$

$$\therefore 0 + b^2 + 0 + 0 + 2bv + 0 + 0 = 0$$
$$\therefore b^2 + 2bv = 0$$
$$\therefore v = \frac{-b}{2}.$$

It passes through $C\,(0, 0, c)$

$$\therefore 0 + 0 + c^2 + 0 + 0 + 2cw + 0 = 0$$
$$\therefore c^2 + 2cw = 0$$
$$\therefore w = \frac{-c}{2}.$$

$\therefore$ Equation (3.15) becomes;

$$x^2 + y^2 + z^2 + 2\left(\tfrac{-a}{2}\right)x + 2\left(\tfrac{-b}{2}\right)y + 2\left(\tfrac{-c}{2}\right)z = 0$$
$$\therefore x^2 + y^2 + z^2 - ax - by - cz = 0$$

is the required equation of a sphere.

8) Obtain the equation of the sphere which passes through the three points (1, 0, 0); (0, 1, 0); (0, 0, 1) and has its radius as small as possible.

Sol. Suppose the sphere passing through three points $A(1, 0, 0)$; $B(0, 1, 0)$; $C(0, 0, 1)$ is

$$x^2 + y^2 + z^2 + 2ux + 2vy + 2wz + d = 0. \qquad (3.16)$$

It passes through $A(1, 0, 0)$
$$\therefore 1^2 + 0 + 0 + 2u + 0 + 0 + d = 0$$
$$\therefore 2u = -d - 1.$$

It passes through $B(0, 1, 0)$
$$\therefore 0 + 1 + 0 + 0 + 2v + 0 + d = 0$$
$$\therefore 2v = -d - 1.$$

It passes through $C(0, 0, 1)$
$$\therefore 0 + 0 + 1 + 0 + 0 + 2w + d = 0$$
$$\therefore 2w = -d - 1$$

Radius $r = \sqrt{u^2 + v^2 + w^2 - d}$
$$= \sqrt{\left(\frac{-d-1}{2}\right)^2 + \left(\frac{-d-1}{2}\right)^2 + \left(\frac{-d-1}{2}\right)^2 - d}$$
$$= \sqrt{\frac{3(d^2+2d+1)-4d}{4}}$$
$$\therefore r = \sqrt{\frac{3(d^2+2d+1)-4d}{4}}$$
$$\therefore r = \frac{3d^2+2d+3}{4}.$$

$\therefore$ Radius will be minimum if derivative of $r = 0$.
$$\text{i.e., } 6d + 2 = 0$$
$$\therefore d = \frac{-1}{3}$$
$$\therefore 2u = \frac{+1}{3} - 1 = \frac{-2}{3}; \ 2v = \frac{+1}{3} - 1 = \frac{-2}{3}; \ 2w = \frac{+1}{3} - 1 = \frac{-2}{3}$$

$\therefore$ The required equation of the sphere is
$$x^2 + y^2 + z^2 - \frac{2}{3}x - \frac{2}{3}y - \frac{2}{3}z - \frac{1}{3} = 0$$
$$\therefore 3\left(x^2 + y^2 + z^2\right) - 2(x + y + z) - 1 = 0.$$

10) Show that the equation of the sphere passing through the three points $(3, 0, 2)$; $(-1, 1, 1)$; $(2, -5, 4)$ and having its center on the plane $2x + 3y + 4z = 6$ is $x^2 + y^2 + z^2 + 4y - 6z = 1$.

Sol. The general equation of sphere is

$$x^2 + y^2 + z^2 + 2ux + 2vy + 2wz + d = 0. \tag{3.17}$$

It passes through $A\,(3, 0, 2)$

$$\therefore 3^2 + 0 + 2^2 + 6u + 0 + 4w + d = 0$$
$$\therefore 6u + 4w + d = -13. \tag{3.18}$$

It passes through $B\,(-1, 1, 1)$

$$\therefore 1 + 1 + 1 + (-2)\,u + 2v + 2w + d = 0$$
$$\therefore -2u + 2v + 2w + d = -3. \tag{3.19}$$

It passes through C$(2, -5, 4)$

$$\therefore 4 + 25 + 16 + 4u - 10v + 8w + d = 0$$
$$\therefore 4u - 10v + 8w + d = -45. \tag{3.20}$$

The Center lies on the plane $2x + 3y + 4z = 6$.
$\therefore$ Center $(-u, -v, -w)$ satisfies the equation of a plane

$$\therefore 2u + 3v + 4w = -6. \tag{3.21}$$

Solving Equations (3.19) and (3.21), we get

$$\therefore 5v + 6w + d = -9. \tag{3.22}$$

Solving Equations (3.18) and (3.22), we get

$$6u + 4w + d = -3 \Rightarrow u = \frac{-13 - d - 4w}{6} = \frac{-1}{6}\,(13 + d + 4w)$$

$$5v + 6w + d = -9 \Rightarrow v = \frac{-9 - d - 6w}{5} = \frac{-1}{5}\,(9 + d + 6w).$$

Substituting the values of u and v in Equation (3.20), we get

$$\therefore 4\left(\tfrac{-1}{6}\,(13 + d + 4w)\right) - 10\left(\tfrac{-1}{5}\,(9 + 6w + d)\right) + 8w + d = -45$$
$$\therefore -26 - 2d - 8w + 54 + 36w + 6d + 24w + 3d = -135$$
$$\therefore 52w + 7d = -163. \tag{3.23}$$

3.5 Equation of a Sphere Passing through the Four Points

Substituting the values of u and v in Equation (3.21), we get

$$\therefore 2\left(\tfrac{-1}{6}(13 + d + 4w)\right) + 3\left(\tfrac{-1}{5}(9 + d + 6w)\right) + 4w = -6$$

$$\therefore -65 - 5d - 20w - 81 - 54w - 9d + 60w = -90$$

$$\therefore -14w - 14d = 56$$

$$\therefore w + d = -4. \tag{3.24}$$

Solving (3.23) and (3.24), we get

$$45w = -135$$

$$\therefore w = -3$$

$$\therefore -3 + d = -4 \Rightarrow d = -1$$

$$\therefore u = \frac{-1}{6}(4w + d + 13) = \frac{-1}{6}(-12 - 1 + 13) = 0$$

$$\therefore v = \frac{-1}{5}(6w + d + 9) = \frac{-1}{5}(-18 - 1 + 9) = 2$$

$\therefore$ The Equation (3.17) becomes, $x^2 + y^2 + z^2 + 4y - 6z = 1$.

11) Obtain the sphere having its center on the line $5y + 2z = 0 = 2x - 3y$ and passing through the two points $(0, -2, -4)$; $(2, -1, -1)$.

Sol. Let the general equation of sphere be

$$x^2 + y^2 + z^2 + 2ux + 2vy + 2wz + d = 0. \tag{3.25}$$

$\therefore$ Center $(-u, -v, -w)$ lies on the line $5y + 2z = 0 = 2x - 3y$

$$\therefore -5v - 2w = 0 \text{ and } -2u + 3v = 0$$

$$\therefore 2w = -5v \text{ and } 2u = 3v.$$

The Equation (3.25) passes through points $(0, -2, -4)$ and $(2, -1, -1)$

$$\therefore 0 + (-2)^2 + (-4)^2 + 0 - 4v - 8w + d = 0$$

$$\therefore -4v + 8w + d - 20 = 0$$

$$\therefore -4v + 20v + d + 20 = 0 \Rightarrow 16v + d = -20 \tag{3.26}$$

and

$$(2)^2 + (-1)^2 + (-1)^2 + 4u - 2v - 2w + d = 0$$

$$\therefore 4u - 2v + 2w + d + 6 = 0$$

$$\therefore 6v - 2v + 5v + d + 6 = 0$$

$$\therefore 9v + d = -6. \tag{3.27}$$

Solving (3.26) and (3.27), we get

$$v = -2$$
$$9v + d = -6$$
$$\therefore 9(-2) + d = -6$$
$$\therefore d = 12$$

But $2w = -5v$ and $2u = 3v$;

$$\therefore w = 5 \text{ and } u = -3.$$

$\therefore$ The Equation (3.25) becomes;

$$x^2 + y^2 + z^2 + 6x - 4y + 10z + 12 = 0.$$

12) A sphere of constant radius r passes through the origin O and cuts the axes in A, B, C. Find the locus of the foot of the perpendicular from O to the plane ABC.

Sol. Suppose the coordinates of the points A, B, C are $(a, 0, 0)$; $(0, b, 0)$; $(0, 0, c)$ respectively.
$\therefore$ The equation of sphere OABC is

$$x^2 + y^2 + z^2 - ax - by - cz = 0. \tag{3.28}$$

The equation of the plane is

$$\frac{x}{a} + \frac{y}{b} + \frac{z}{c} = 1 \tag{3.29}$$

The equation of the line through the origin perpendicular to this plane are

$$\frac{x}{\frac{1}{a}} = \frac{y}{\frac{1}{b}} = \frac{z}{\frac{1}{c}}. \tag{3.30}$$

Let $P(x_1, y_1, z_1)$ be the coordinates of foot of perpendicular which satisfies Equations (3.29) and (3.30)

$$\therefore \frac{x_1}{a} + \frac{y_1}{b} + \frac{z_1}{c} = 1. \tag{3.31}$$

$$\therefore \frac{x_1}{\frac{1}{a}} = \frac{y_1}{\frac{1}{b}} = \frac{z_1}{\frac{1}{c}} = \lambda$$

3.5 Equation of a Sphere Passing through the Four Points

$$\therefore a = \frac{\lambda}{x_1}; b = \frac{\lambda}{y_1}; c = \frac{\lambda}{z_1}.$$

The radius r of the sphere is

$$r^2 = \left(\frac{a}{2}\right)^2 + \left(\frac{b}{2}\right)^2 + \left(\frac{c}{2}\right)^2$$
$$\therefore a^2 + b^2 + c^2 = 4r^2$$
$$\therefore \lambda^2 \left(\frac{1}{x_1^2} + \frac{1}{y_1^2} + \frac{1}{z_1^2}\right) = 4r^2. \tag{3.32}$$

Substituting the values of a, b, and c in Equation (3.31), we get

$$\therefore \frac{x_1}{\frac{\lambda}{x_1}} + \frac{y_1}{\frac{\lambda}{y_1}} + \frac{z_1}{\frac{\lambda}{z_1}} = 1$$
$$\therefore \frac{1}{\lambda}\left(x_1^2 + y_1^2 + z_1^2\right) = 1$$
$$\therefore \frac{1}{\lambda^2}\left(x_1^2 + y_1^2 + z_1^2\right)^2 = 1. \tag{3.33}$$

Multiplying Equations (3.32) and (3.33), we get

$$\therefore \left(x_1^2 + y_1^2 + z_1^2\right)^2 \left(\frac{1}{x_1^2} + \frac{1}{y_1^2} + \frac{1}{z_1^2}\right) = 4r^2$$

$$\therefore \left(x^2 + y^2 + z^2\right)^2 \left(\frac{1}{x^2} + \frac{1}{y^2} + \frac{1}{z^2}\right) = 4r^2$$

which is the required locus.

13) If O be the center of a sphere of radius unity and A, B be two points in a line with O such that $OA.OB = 1$ and if P be a variable point on the sphere, show that $PA : PB = $ constant.

Sol. Let O be the origin and line OAB be the x axis.
$\therefore$ Coordinate of A are $(a, 0, 0)$ and coordinate of B are $\left(\frac{1}{a}, 0, 0\right)$.

$$\therefore OA \cdot OB = 1$$

$\therefore$ The equation of sphere with center origin and radius unity is

$$x^2 + y^2 + z^2 = 1.$$

Let $P(x_1, y_1, z_1)$ be any point of the sphere

$$\therefore x_1^2 + y_1^2 + z_1^2 = 1$$

$$\therefore \frac{PA}{PB} = \frac{\sqrt{(x_1-a)^2 + y_1^2 + z_1^2}}{\sqrt{(x_1-\frac{1}{a})^2 + y_1^2 + z_1^2}}$$

$$= \frac{\sqrt{x_1^2 + y_1^2 + z_1^2 - 2ax_1 + a^2}}{\sqrt{x_1^2 + y_1^2 + z_1^2 - \frac{2x_1}{a} + \frac{1}{a^2}}}$$

$$= \frac{\sqrt{1 - 2ax_1 + a^2}}{\sqrt{1 - \frac{2x_1}{a} + \frac{1}{a^2}}}$$

$$= \frac{a.\sqrt{1 - 2ax_1 + a^2}}{\sqrt{1 - 2ax_1 + a^2}}$$

$$= a = \text{constant}.$$

14) A sphere of constant radius $2k$ passes through the origin and meets the axes in A, B, C. Show that the locus of the centroid of the tetrahedron OA BC is the sphere $x^2 + y^2 + z^2 = k^2$.

Sol. Suppose the coordinates of points A, B, C are $(a, 0, 0)$; $(0, b, 0)$; $(0, 0, c)$ respectively.

$\therefore$ The equation of sphere OA BC is $x^2 + y^2 + z^2 - ax - by - cz = 0$.

$\therefore$ Radius $r = \sqrt{\left(\frac{a}{2}\right)^2 + \left(\frac{b}{2}\right)^2 + \left(\frac{c}{2}\right)^2} = \sqrt{\frac{a^2+b^2+c^2}{4}}$.

Given $r = 2k$

$$\therefore 2k = \sqrt{\frac{a^2 + b^2 + c^2}{4}} \tag{3.34}$$

$$\therefore 16k^2 = a^2 + b^2 + c^2$$

Let $P(x_1, y_1, z_1)$ be the centroid of the tetrahedron OA BC then

$$x_1 = \frac{a}{4};\ y_1 = \frac{b}{4};\ z_1 = \frac{c}{4}$$

$$\therefore a = 4x_1;\ b = 4y_1;\ c = 4z_1$$

$\therefore$ The Equation (3.34) becomes;

$$16k^2 = 16x_1^2 + 16y_1^2 + 16z_1^2$$

$$\therefore x_1^2 + y_1^2 + z_1^2 = k^2$$

$$\therefore x^2 + y^2 + z^2 = k^2.$$

3.6 Section of the Sphere by a Plane

Consider a sphere with a center at C and radius a. Let A be a point on the section of the sphere by the given plane. Let CP be the perpendicular distance of the plane from the center C of the sphere. Since CP is perpendicular to the plane.

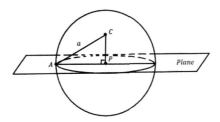

$\therefore$ It is perpendicular to the line AP.

$$\therefore \angle CPA = 90$$
$$\therefore CP^2 + AP^2 = AC^2 = a^2 \; (\because AC \text{ is radius of sphere})$$
$$\therefore AP^2 = a^2 - CP^2 = \text{constant}$$

$\therefore$ the Locus of A is a circle with a center at P and radius AP.

$\therefore$ The plane section is a circle.

Remark:

1) If $CP > a$; i.e., AP^2 is negative, then the circle is imaginary.

2) $x^2 + y^2 + z^2 + 2ux + 2vy + 2wz + d = 0$ and $ax + by + cz + k = 0$ taken together represents a circle.

3) The section of the sphere by a plane passing through its center is called a great circle.

4) The center and radius of the great circle are the same as the center and radius of the sphere.

5) The equation
$$\left(x^2 + y^2 + z^2 + 2ux + 2vy + 2wz + d\right) + \lambda\left(ax + by + cz + k\right) = 0$$
represents a sphere passing through the circle
$$\begin{cases} x^2 + y^2 + z^2 + 2ux + 2vy + 2wz + d = 0 \\ ax + by + cz + k = 0. \end{cases}$$

3.7 Intersection of Two Spheres

Consider two spheres whose equations are

$$S_1 : x^2 + y^2 + z^2 + 2u_1 x + 2v_1 y + 2w_1 z + d_1 = 0,$$
$$S_2 : x^2 + y^2 + z^2 + 2u_2 x + 2v_2 y + 2w_2 z + d_2 = 0.$$

Any point common to both the sphere; satisfies both the equation

$$S_1 = 0 \text{ and } S_2 = 0.$$

$\therefore$ It satisfies

$$S_1 - S_2 = 2(u_1 - u_2)x + 2(v_1 - v_2)y + 2(w_1 - w_2)z + d_1 - d_2 = 0$$

which is a linear equation in x, y, z.
$\therefore$ It is the equation of a plane.

Remark:

The points of intersections of two spheres $S_1 = 0$ and $S_2 = 0$ are the same as the points of intersection of one of two spheres and the plane $S_1 - S_2 = 0$; ; but the intersection of sphere and plane is a circle.
$\therefore$ The intersection of two spheres $S_1 = 0$ and $S_2 = 0$ is a circle.

15) Find the center and the radiu the circle

$$x + 2y + 2z = 15; \quad x^2 + y^2 + z^2 - 4z - 2y = 11.$$

Sol. Given the equation of the circle is

$$S : x^2 + y^2 + z^2 - 2y - 4z - 11 = 0$$
$$P : x + 2y + 2z - 15 = 0.$$

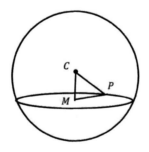

The center of the sphere is $C(0, 1, 2)$ and $r = \sqrt{0 + 1 + 4 + 11} = 4$

3.7 Intersection of Two Spheres

The length of the perpendicular from the center $(0, 1, 2)$ to the plane is

$$p = \left|\frac{0(1) + 2(1) + 2(2) - 15}{\sqrt{1^2 + 2^2 + 2^2}}\right| = \left|\frac{-9}{3}\right| = 3$$

$\therefore CP = 4$ and $CM = 3$.

$\therefore$ The radius of the circle $= PM = \sqrt{CP^2 - CM^2} = \sqrt{16 - 9} = \sqrt{7}$.

Equation of line through the center C and perpendicular to the given plane is

$$\frac{x-0}{1} = \frac{y-1}{2} = \frac{z-2}{2} = k$$

$\therefore x = k;\ y = 2k + 1;\ z = 2k + 2$.

Coordinates of line satisfy the equation of the plane

i.e., $x + 2y + 2z - 15 = 0$

$\therefore k + 2(2k+1) + 2(2k+2) - 15 = 0$

$\therefore k + 4k + 2 + 4k + 4 - 15 = 0$

$\therefore 9k - 9 = 0$

$\therefore k = 1$

$\therefore$ The center of the circle is $(1,\ 2(1) + 1,\ 2(2) + 2) = (1, 3, 6)$.

16) Find the equation of the circle lying on the sphere $x^2 + y^2 + z^2 - 2x + 4y - 6z + 3 = 0$ and having its center at $(2, 3, -4)$.

Sol. The given equation of sphere is

$$x^2 + y^2 + z^2 - 2x + 4y - 6z + 3 = 0.$$

$\therefore$ Comparing with the general equation

$$x^2 + y^2 + z^2 + 2ux + 2vy + 2wz + d = 0$$

$\therefore 2u = -2;\ 2v = 4;\ 2w = -6;\ d = 3$

$\therefore u = -1;\ v = 2;\ w = -3;\ d = 3$

$\therefore$ The center of the sphere is $C(1, -2, 3)$, and the center of the circle is $B(2, 3, -4)$.

$\therefore$ Direction ratios of line BC are $(1, 5, -7)$.

∴ The plane through the point $(2,\ 3,\ -4)$ and direction ratios $(1,\ 5,\ -7)$ is

$$1(x-2) + 5(y-3) - 7(z+4) = 0$$

$$\therefore x + 5y - 7z - 45 = 0.$$

∴ The equation of the circle is

$$S: x^2 + y^2 + z^2 - 2x + 4y - 6z + 3 = 0$$

$$P: x + 5y - 7z - 45 = 0.$$

17) Find the equation of the sphere through the circle $x^2 + y^2 + z^2 = 9;\ 2x + 3y + 4z = 5$ and the point $(1,\ 2,\ 3)$.

Sol. The equation of sphere is

$$S + \lambda P = 0$$

$$\therefore (x^2 + y^2 + z^2 - 9) + \lambda(2x + 3y + 4z - 5) = 0 \quad (3.35)$$

It passes through the point $(1,\ 2,\ 3)$

$$\therefore (1 + 4 + 9 - 9) + \lambda(2 + 6 + 12 - 5) = 0$$

$$\therefore 5 + 15\lambda = 0$$

$$\therefore \lambda = \frac{-1}{3}$$

∴ The Equation (3.35) becomes;

$$(x^2 + y^2 + z^2 - 9) \frac{-1}{3}(2x + 3y + 4z - 5) = 0$$

$$\therefore 3x^2 + 3y^2 + 3z^2 - 2x - 3y - 4z - 22 = 0.$$

18) Show that the two circles

$$x^2 + y^2 + z^2 - y + 2z = 0,\ x - y + z - 2 = 0 \text{ and}$$

$$x^2 + y^2 + z^2 + x - 3y + z - 5 = 0,\ 2x - y + 4z - 1 = 0$$

lie on the same sphere and find its equation.

Sol. The equation of sphere is $S + \lambda P = 0$.

$$\therefore x^2 + y^2 + z^2 - y + 2z + \lambda_1(x - y + z - 2) = 0 \quad (3.36)$$

$$x^2 + y^2 + z^2 + x - 3y + z - 5 + \lambda_2 (2x - y + 4z - 1) = 0). \quad (3.37)$$

Equations (3.36) and (3.37) represents the same sphere, so comparing coefficient; we get

$$\lambda_1 = 2\lambda_2 + 1; \quad -\lambda_1 - 1 = -3 - \lambda_2; \quad 2 + \lambda_1 = 1 + 4\lambda_2$$

$$\therefore \lambda_1 - 2\lambda_2 = 1; \quad -\lambda_1 + \lambda_2 = 2$$

$$\therefore \lambda_2 = 1 \text{ and } \lambda_1 = 3$$

Substituting the value of λ_1 in equation (3.36), we get

$$\therefore x^2 + y^2 + z^2 - y + 2z + 3(x - y + z - 2) = 0$$

$$\therefore x^2 + y^2 + z^2 + 3x - 4y + 5z - 6 = 0.$$

19) Find the equation of the sphere through the circle $x^2 + y^2 + z^2 + 2x + 3y + 6 = 0; \, ; x - 2y + 4z - 9 = 0$ and the center of the sphere $x^2 + y^2 + z^2 - 2x + 4y - 6z + 5 = 0$.

Sol. The equation of the circle is

$$S : x^2 + y^2 + z^2 + 2x + 3y + 6 = 0$$

$$P : x - 2y + 4z - 9 = 0$$

∴ The equation of the circle is

$$S + \lambda P = 0$$

$$\therefore (x^2 + y^2 + z^2 + 2x + 3y + 6) + \lambda (x - 2y + 4z - 9) = 0 \quad (3.38)$$

The given equation of sphere is $x^2 + y^2 + z^2 - 2x + 4y - 6z + 5 = 0$.
∴ Center $C = (1, -2, 3)$
Sphere (3.38) passes through center $C = (1, -2, 3)$.
∴ It satisfies the equation of the sphere

$$\therefore \left[(1)^2 + (-2)^2 + (3)^2 + 2(1) + 3(-2) + 6 \right]$$

$$+ \lambda [1 - 2(-2) + 4(3) - 9] = 0$$

$$\therefore 16 + 8\lambda = 0$$

$$\therefore \lambda = -2$$

∴ Equation (3.38) becomes;

$$(x^2 + y^2 + z^2 + 2x + 3y + 6) - 2(x - 2y - 4z - 9) = 0$$

$$\therefore x^2 + y^2 + z^2 + 7y - 8z + 24 = 0.$$

20) Show that the equation of the sphere having its center on the plane $4x - 5y - z = 3$ and passing through the circle with equation $x^2 + y^2 + z^2 - 2x - 3y + 4z + 8 = 0$; $x^2 + y^2 + z^2 + 4x + 5y - 6z + 2 = 0$ is $x^2 + y^2 + z^2 + 7x + 9y - 11z - 1 = 0$.

Sol. The given circle is the intersection of spheres

$$S_1 : x^2 + y^2 + z^2 - 2x - 3y + 4z + 8 = 0, \qquad (3.39)$$

$$S_2 : x^2 + y^2 + z^2 + 4x + 5y - 6z + 2 = 0. \qquad (3.40)$$

The equation of the plane of the circle is $S_1 - S_2 = 0$

$$\therefore -6x - 8y + 10z + 6 = 0$$
$$\therefore 3x + 4y - 5z - 3 = 0. \qquad (3.41)$$

The equation of sphere is $S + \lambda P = 0$

$$\therefore (x^2 + y^2 + z^2 - 2x - 3y + 4z + 8) + \lambda(3x + 4y - 5z - 3) = 0$$

$$\therefore x^2 + y^2 + z^2 + (-2 + 3\lambda)x + (-3 + 4\lambda)y + (4 - 5\lambda)z$$
$$+ (8 - 2\lambda) = 0$$

∴ The center of the sphere is $\left(\frac{2-3\lambda}{2}; \frac{3-4\lambda}{2}; ; \frac{-4+5\lambda}{2}\right)$
But given that center lies on the plane $4x - 5y - z = 3$

$$\therefore 4\left(\frac{2-3\lambda}{2}\right) - 5\left(\frac{3-4\lambda}{2}\right) - \left(\frac{-4+5\lambda}{2}\right) = 3$$

$$\therefore 8 - 12\lambda - 15 + 20\lambda + 4 - 5\lambda = 6$$

$$\therefore 3\lambda = 9$$

$$\therefore \lambda = 3.$$

Substituting the value of λ, the equation becomes,

$$\therefore (x^2 + y^2 + z^2 - 2x - 3y + 4z + 8) + 3(3x + 4y + 5z - 3) = 0$$

3.7 Intersection of Two Spheres

$$\therefore x^2 + y^2 + z^2 + 7x + 9y - 11z - 1 = 0.$$

21) Obtain the equation of the sphere having the circle $x^2 + y^2 + z^2 + 10y - 4z - 8 = 0$; $x + y + z = 3$ as a great circle.

Sol. The given equation of the circle is

$$S : x^2 + y^2 + z^2 + 10y - 4z - 8 = 0 \qquad (3.42)$$

$$P : x + y + z - 3 = 0. \qquad (3.43)$$

The equation of sphere is $S + \lambda P = 0$

$$\therefore (x^2 + y^2 + z^3 + 10y - 4z - 8) + \lambda(x + y + z - 3) = 0 \qquad (3.44)$$

$$\therefore x^2 + y^2 + z^2 + \lambda x + (10 + \lambda)y + (-4 + \lambda)z - 8 - 3\lambda = 0$$

$\therefore$ The center of the sphere is $\left[\frac{-\lambda}{2}; \frac{-10-\lambda}{2}; \frac{4-\lambda}{2}\right]$

The given circle will be the great circle of (3.44) if the center of the sphere lies on the plane (3.43).

$\therefore$ Center satisfies the equation of the plane

$$\frac{-\lambda}{2} - \left(\frac{10+\lambda}{2}\right) + \left(\frac{4-\lambda}{2}\right) - 3 = 0$$

$$\therefore -\lambda - 10 - \lambda + 4 - \lambda - 6 = 0$$

$$\therefore -3\lambda = 12$$

$$\therefore \lambda = -4$$

$\therefore$ Equation (3.44) becomes;

$$(x^2 + y^2 + z^2 + 10y - 4z - 8) + (-4)(x + y + z - 3) = 0$$

$$\therefore x^2 + y^2 + z^2 - 4x + 6y - 8z + 4 = 0.$$

22) A sphere S has points $(0, 1, 0), (3, -5, 2)$ at opposite ends of a diameter. Find the equation of the sphere having the intersection of the sphere S with the plane $5x - 2y + 4z + 7 = 0$ as a great circle.

Sol. The equation of sphere whose diameter opposite ends are given is

$$(x - x_1)(x - x_2) + (y - y_1)(y - y_2) + (z - z_1) + (z - z_2) = 0$$

$$\therefore (x - 0)(x - 3) + (y - 1)(y + 5) + (z - 0)(z - 2) = 0$$

$$\therefore x^2 - 3x + y^2 + 4y - 5 + z^2 - 2z = 0$$
$$\therefore x^2 + y^2 + z^2 - 3x + 4y - 2z - 5 = 0. \qquad (3.45)$$

The equation of the plane is
$$5x - 2y + 4z + 7 = 0 \qquad (3.46)$$

∴ The equation of sphere is
$$\left(x^2 + y^2 + z^2 - 3x + 4y + 2z - 5\right) + \lambda\left(5x - 2y + 4z + 7\right) = 0$$
$$\therefore x^2 + y^2 + z^2 + (-3 + 5\lambda)x + (4 - 2\lambda)y + (-2 + 4\lambda)z$$
$$+ (-5 + 7\lambda) = 0 \qquad (3.47)$$

∴ Center is $\left[\frac{3-5\lambda}{2}, \frac{2\lambda-4}{2}, \frac{2-4\lambda}{2}\right]$

The intersection of the spheres (3.45) and (3.46) will be a great circle of the sphere (3.47) if the center of (3.47) lies on the plane (3.46).
i.e. the center satisfies the equation of the sphere

$$\therefore 5\left(\frac{3-5\lambda}{2}\right) - 2\left(\frac{2\lambda-4}{2}\right) + 4\left(\frac{2-4\lambda}{2}\right) + 7 = 0$$
$$\therefore 15 - 25\lambda - 4\lambda + 8 + 8 - 16\lambda + 14 = 0$$
$$\therefore -45\lambda + 45 = 0$$
$$\therefore \lambda = 1$$

∴ Substituting the value of λ in Equation (3.47); we get
$$x^2 + y^2 + z^2 + 2x + 2y + 2z + 2 = 0.$$

23) Obtain the equation of the sphere which passes through the circle $x^2 + y^2 = 4; z = 0$ and is cut by the plane $x + 2y + 2z = 0$ in a circle of radius 3.

Sol. Given circle $x^2 + y^2 = 4$ and $z = 0$ is the intersection of cylinder and plane.

∴ Generators of the cylinder are parallel to the z axis.

∴ The center of the circle is the origin and so the center of the sphere lies on the z-axis.

Let $(0, 0, k)$ be the center of the sphere.
∴ The radius of the sphere is $\sqrt{0 + 0 + k^2 + 4} = \sqrt{4 + k^2}$

∴ Equation of sphere through circle is
$$(x-0)^2 + (y-0)^2 + (z-k)^2 = 4 + k^2$$
$$\therefore S: x^2 + y^2 + z^2 - 2kz - 4 = 0, \qquad (3.48)$$
and
$$P: x + 2y + 2z = 0. \qquad (3.49)$$

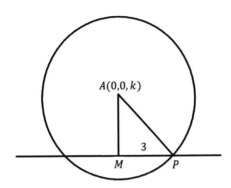

AM= length of the perpendicular from $(0,\ 0,\ k)$ on the plane
$$\therefore AM = \frac{0(1) + 0(2) + 0(2)}{\sqrt{1 + 2^2 + 2^2}} = \frac{2k}{3}$$
In $\triangle AMP$; $AP^2 = AM^2 + MP^2$
$$\therefore k^2 + 4 = \frac{4k^2}{9} + 9$$
$$\therefore 5k^2 = 45$$
$$\therefore k = \pm 3$$
Substituting the value of k in Equation (3.48), we get
$$x^2 + y^2 + z^2 \pm 6z - 4 = 0.$$

24) Show that the two circles
$$2(x^2 + y^2 + z^2) + 8x - 13y + 17z - 17 = 0;\ 2x + y - 3z + 1 = 0$$

and $x^2 + y^2 + z^2 + 3x - 4y + 3z = 0$; $x - y + 2z - 4 = 0$
lie on the same sphere and find its equation.

Sol. Given the equation of circle as

$$2(x^2 + y^2 + z^2) + 8x - 13y + 17z - 17 = 0; 2x + y - 3z + 1 = 0$$

$$\therefore S_1 : (2(x^2 + y^2 + z^2) + 8x - 13y + 17z - 17)$$
$$+ \lambda_1 (2x + y - 3z + 1) = 0 \tag{3.50}$$

$$x^2 + y^2 + z^2 + 3x - 4y + 3z = 0; x - y + 2x - 4 = 0$$

$$S_2 : (x^2 + y^2 + z^2 + 3x - 4y + 3z)$$
$$+ \lambda_2 (x - y + 2z - 4) = 0. \tag{3.51}$$

Two given spheres will lie on the same sphere if for the same values of λ_1 and λ_2 the two spheres are identical.

$\therefore$ The Equation (3.50) becomes;

$$x^2 + y^2 + z^2 + 4x - \frac{13}{2}y + \frac{17}{2}z - \frac{17}{2} + \frac{\lambda_1}{2}(2x + y - 3z + 1) = 0$$

$$\therefore x^2 + y^2 + z^2 + (4 + \lambda_1)x + \left(\frac{-13}{2} + \frac{\lambda_1}{2}\right)y + \left(\frac{17}{2} - \frac{3\lambda_1}{2}\right)z$$
$$+ \left(\frac{-17}{2} + \frac{\lambda_1}{2}\right) = 0.$$

$\therefore$ The Equation (3.51) becomes;

$$x^2 + y^2 + z^2 + (3 + \lambda_2)x + (-4 - \lambda_2)y + (3 + 2\lambda_2)z + (-4\lambda_2) = 0$$

Comparing coefficients, we get

$$4 + \lambda_1 = 3 + \lambda_2; \frac{-13}{2} + \frac{\lambda_1}{2} = -4 - \lambda_2$$

$$\therefore \lambda_1 - \lambda_2 = -1 \frac{\lambda_1}{2} + \lambda_2 = -4 + \frac{13}{2}$$

$$\therefore \lambda_2 = 2 \text{ and } \lambda_1 = 1.$$

$\therefore$ The Equation (3.50) becomes;

$$[2(x^2 + y^2 + z^2) + 8x - 13y + 17z - 17] + 1[2x_1 + y - 3z + 1] = 0$$

$$\therefore 2x^2 + 2y^2 + 2z^2 + 10x - 12y + 14z - 16 = 0$$
$$\therefore x^2 + y^2 + z^2 + 5x - 6y + 7z - 8 = 0$$

∴ The Equation (3.51) becomes;

$$\left(x^2 + y^2 + z^2 + 3x - 4y + 3z\right) + (2)(x + y + 2z - 4) = 0$$
$$\therefore x^2 + y^2 + z^2 + 5x - 6y - 7x - 8 = 0$$

∴ Two circles lie on the same sphere.

3.8 Intersection of Sphere S and Line L

Let the equation of sphere be

$$x^2 + y^2 + z^2 + 2ux + 2vy + 2wz + d = 0, \qquad (3.52)$$

and equation of the line be

$$\frac{x - \alpha}{l} = \frac{y - \beta}{m} = \frac{z - \gamma}{n} = r \qquad (3.53)$$

∴ The coordinates of the line (3.53) are $x = rl + \alpha$; $y = rm + \beta$; $z = rn + \gamma$. The line intersects the sphere.

∴ Coordinate of line satisfy the equation of sphere (3.52).

$$\therefore (\alpha + lr)^2 + (\beta + mr)^2 + (\gamma + nr)^2 + 2u(\alpha + lr) + 2v(\beta + mr)$$
$$+ 2w(\gamma + nr) + d = 0$$
$$\therefore \left(l^2 + m^2 + n^2\right)r^2 + 2r(l\alpha + m\beta + n\gamma + ul + vm + wn)$$
$$+ \left(\alpha^2 + \beta^2 + \gamma^2 + 2u\alpha + 2v\beta + 2w\gamma + d\right) = 0. \qquad (3.54)$$

∴ Equation (3.54) is a quadratic equation in r.
∴ Line l meet sphere in two points if $\Delta > 0$.
Here, $a = l^2 + m^2 + n^2$; $b = 2(l\alpha + m\beta + n\gamma + ul + vm + nw)$ and $c = \alpha^2 + \beta^2 + \gamma^2 + 2u\alpha + 2v\beta + 2w\gamma + d$
∴ By Equation (3.54); points of contact are

$$P(\alpha + lr_1; \beta + mr_2; \gamma + nr_1) \text{ and } Q(\alpha + lr_2; \beta + mr_2; \gamma + nr_2)$$

If $\Delta = 0$; i.e., $b^2 - 4ac = 0 \therefore b^2 = 4ac$

∴ The line intersects the sphere in two coincident points.
∴ The roots of the Equation (3.54) are r_1 and r_2 which are equal.
i.e., the line intersects the sphere at only one point. So, the line will become a tangent.
If $\Delta < 0$; the line does not intersect the sphere.

3.9 Tangent Plane

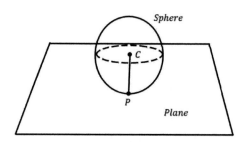

The locus of all tangent lines at a point P of the sphere is called the tangent plane to the sphere at a point P.

A plane that touches the given sphere exactly at one point is called a tangent plane to the sphere at that point.

Consider a sphere with center C and a plane that touches it at P then a plane is a tangent plane to the sphere.

i.e., the length of the perpendicular from the center C of the sphere to the tangent plane is equal to the radius of the sphere.

Equation of the tangent plane:

The general equation of sphere is

$$x^2 + y^2 + z^2 + 2ux + 2vy + 2wz + d = 0$$

and a tangent plane intersects the sphere at $P(x_1, y_1, z_1)$.

$\therefore$ The center of the sphere is $C(-u, -v, -w)$.

Direction ratios of the normal CP to the plane are $(x_1 + u;\ y_1 + v;\ z_1 + w)$.

$\therefore$ Equation of plane through $P(x_1, y_1, z_1)$, and normal to CP is

$$(x_1 + u)(x - x_1) + (y_1 + v)(y - y_1) + (z_1 + w)(z - z_1) = 0$$

$$\therefore xx_1 - x_1^2 + ux - ux_1 + yy_1 - y_1^2 + vy - vy_1 + zz_1 - z_1^2$$
$$+ wz - wz_1 = 0$$

$\therefore xx_1 + yy_1 + zz_1 + ux + vy + wz = (x_1^2 + y_1^2 + z_1^2 + 2ux_1 + 2vy_1 + 2wz_1)$
$- ux_1 - vy_1 - wz_1 = -ux_1 - vy_1 - wz_1$.

$\therefore$ The equation of the tangent plane is

$$xx_1 + yy_1 + zz_1 + u(x + x_1) + v(y + y_1) + w(z + z_1) + d = 0.$$

Remark:

1) The equation of the tangent plane to be the sphere $x^2 + y^2 + z^2 = a^2$ at point $P(x_1, y_1, z_1)$ is $xx_1 + yy_1 + zz_1 = a^2$

2) The equation of the tangent plane of the sphere can be written by replacing x^2 by xx_1; y^2 by yy_1; z^2 by zz_1; $2x$ by $x + x_1$; $2y$ by $y + y_1$ and $2z$ by $z + z_1$ in the general equation of a sphere.

3.10 Equation of the Normal to the Sphere

The line CP which passes through the center C of the sphere and through the point of the contact P and perpendicular to the tangent plane is called normal to the sphere at P.

If l, m, n are the coefficients of x, y, z in the equation of the tangent plane then l, m, n are the direction ratios of the normal. Since it passes through $P(x_1, y_1, z_1)$ its equation is $\frac{x-x_1}{l} = \frac{y-y_1}{m} = \frac{z-z_1}{n}$.

26) Find the two tangent planes to the sphere $x^2 + y^2 + z^2 - 4x + 2y - 6z + 5 = 0$ which are parallel to the plane $2x + 2y = z$.

Sol. The general equation of a plane parallel to the given plane
$2x + 2y = z$ is $2x + 2y - z + k = 0$.
The plane will be a tangent plane if its distance from the center $(2, -1, 3)$ of the sphere is equal to the radius of sphere $= 3$.
i.e., $p = r$
$$\frac{2(2) + 2(-1) + 3(-1) + k}{\sqrt{4 + 4 + 1}} = 3$$
$$\therefore \frac{k - 1}{\pm 3} = 3$$
$$\therefore k - 1 = \pm 9$$

when $k = 1 + 9$ $\therefore k = 10$ and when $k = 1 - 9$ $\therefore k = -8$
$\therefore$ The required equations of tangent planes are $2x + 2y - z + 10 = 0$ and $2x + 2y - z - 8 = 0$.

27) Find the equation of the sphere which touches the sphere $x^2 + y^2 + z^2 - x + 3y + 2z - 3 = 0$ at the point $(1, 1, -1)$ and passes through the origin.

Sol. The equation of tangent plane to the given sphere at $(1, 1, -1)$ is

$$xx_1 + yy_1 + zz_1 + u(x + x_1) + v(y + y_1) + w(z + z_1) + d = 0. \quad (3.55)$$

Sphere

The equation of sphere is $x^2 + y^2 + z^2 - x + 3y + 2z - 3 = 0$

$\therefore 2u = -1; 2v = 3; 2w = 2; d = -3$

$\therefore u = \dfrac{-1}{2}; v = \dfrac{3}{2}; w = 1.$

$\therefore$ Equation (3.55) becomes

$$x + y - z - \dfrac{1}{2}(x+1) + \dfrac{3}{2}(y+1) + 1(z-1) - 3 = 0$$

$\therefore 2x + 2y + 2z - x - 1 + 3y + 3 + 2z - 2 - 6 = 0$

$\therefore x + 5y - 6 = 0.$

$\therefore$ The required equation of the sphere is $S + \lambda P = 0$.

$\therefore (x^2 + y^2 + z^2 - x + 3y + 2z - 3) + k(x + 5y - 6) = 0$

which passes through the origin

$\therefore -3 - 6k = 0$

$\therefore k = \dfrac{-1}{2}$

$\therefore$ Equation becomes,

$$(x^2 + y^2 + z^2 - x + 3y + 2z - 3)\dfrac{-1}{2}(x + 5y - 6) = 0$$

$\therefore 2x^2 + 2y^2 + 2z^2 - 3x + y + 4z = 0.$

28) Find the equation of the sphere through the circle $x^2 + y^2 + z^2 = 1$; $2x + 4y + 5z = 6$ and touching plane $z = 0$.

Sol. The equation of sphere is $S + \lambda P = 0$

$\therefore (x^2 + y^2 + z^2 - 1) + \lambda(2x + 4y + 5z - 6) = 0$

$\therefore x^2 + y^2 + z^2 + 2\lambda x + 4\lambda y + 5\lambda z - 1 - 6\lambda = 0$

$\therefore 2u = 2\lambda; 2v = 4\lambda; 2w = 5\lambda$

$\therefore u = \lambda; v = 2\lambda; w = \dfrac{5}{2}\lambda; d = -1 - 6\lambda$

$\therefore$ Center is $\left(-\lambda, -2\lambda, \dfrac{-5}{2}\lambda\right)$

3.10 Equation of the Normal to the Sphere

and radius $= \sqrt{\lambda^2 + 4\lambda^2 + \frac{25}{4}\lambda^2 + 1 + 6\lambda} = \sqrt{5\lambda^2 + \frac{25}{4}\lambda^2 + 1 + 6\lambda}$.

Since it touches $z = 0$, the length of the perpendicular from the center to the plane is equal to its radius.

$$\therefore \frac{5}{2}\lambda = \pm\sqrt{5\lambda^2 + \frac{25}{4}\lambda^2 + 1 + 6\lambda}$$

$$\therefore 5\lambda^2 + 6\lambda + 1 = 0$$

$$\therefore (\lambda + 1)(5\lambda + 1) = 0$$

$$\therefore \lambda = -1 \text{ or } \lambda = \frac{-1}{5}$$

$\therefore$ The equation of spheres are $x^2 + y^2 + z^2 - 2x - 4y - 5z + 5 = 0$ and $5x^2 + 5y^2 + 5z^2 - 2x - 4y - 5z + 1 = 0$.

29) Find the equation of the two tangent planes to the sphere $x^2 + y^2 + z^2 = 9$ which passes through the line $x + y = 6; x - 2z = 3$.

Sol. The equation of the plane is $l_1 + \lambda l_2 = 0$.

$$\therefore (x + y - 6) + \lambda(x - 2z - 3) = 0$$

$$\therefore (1 + \lambda)x + y - 2\lambda z - 6 - 3\lambda = 0.$$

Given plane will touch the given sphere if $r = p$

$$\therefore 3 = \frac{-6 - 3\lambda}{\sqrt{(1 + \lambda)^2 + 1 + 4\lambda^2}}$$

$$\therefore (1 + \lambda)^2 + 1 + 4\lambda^2 = (-2 - \lambda)^2$$

$$\therefore 4\lambda^2 - 2\lambda - 2 = 0$$

$$\therefore (\lambda - 1)(2\lambda + 1) = 0$$

$$\therefore \lambda = 1 \text{ or } \lambda = \frac{-1}{2}.$$

$\therefore$ Equation of tangent planes are
$2x + y - 2z - 9 = 0$ and $x + 2y + 2z - 9 = 0$.

30) Find the equation of the tangent plane to the sphere $3(x^2 + y^2 + z^2) - 2x - 3y - 4z - 22 = 0$ at the point $(1, 2, 3)$.

Sol. The equation of the sphere is

$$3\left(x^2+y^2+z^2\right)-2x-3y-4z-22=0$$

$$\therefore x^2+y^2+z^2-\frac{2}{3}x-y-\frac{4}{3}z-\frac{22}{3}=0.$$

$\therefore$ Comparing with the general equation, we get

$$x^2+y^2+z^2+2ux+2vy+2wz+d=0$$

$$\therefore 2u=\frac{-2}{3}; 2v=-1; 2w=\frac{-4}{3}; d=\frac{-22}{3}$$

$$\therefore u=\frac{-1}{3}; v=\frac{-1}{2}; w=\frac{-2}{3}$$

$\therefore$ Center is $C\left(\frac{1}{3},\frac{1}{2},\frac{2}{3}\right)$.

The equation of the tangent plane is

$$xx_1+yy_1+zz_1+u(x+x_1)+v(y+y_1)+w(z+z_1)+d=0$$

$$\therefore x+2y+3z\frac{-1}{3}(x+1)-\frac{1}{2}(y+2)\frac{-2}{3}(z+3)\frac{-22}{3}=0$$

$$\therefore 6x+12y+18z-2x-2-3y-6-4z-12-44=0$$

$$\therefore 4x+9y+14z-64=0$$

is the required equation of the tangent plane.

31) Find the equations of the tangent line to the circle $x^2+y^2+z^2+5x-7y-2z-8=0$; $3x-2y+4z+3=0$ at point $(-3, 5, 4)$.

Sol. The given equation of the circle is

$$S:x^2+y^2+z^2+5x-7y+2z-8=0$$

$$P:3x-2y+4z+3=0.$$

The required line will be the intersection of the given plane of the circle and the tangent plane to the given sphere at the point $(-3, 5, 4)$.

$$\therefore x^2+y^2+z^2+5x-7y+2z-8=0$$

$$\therefore 2u=5; 2v=7; 2w=2; d=-8$$

$$\therefore u=\frac{5}{2}; v=\frac{-7}{2}; w=1.$$

∴ The equation of the tangent plane is

$$xx_1 + yy_1 + zz_1 + u(x + x_1) + v(y + y_1) + w(z + z_1) + d = 0$$

$$\therefore -3x + 5y + 4z + \frac{5}{2}(x - 3) - \frac{7}{2}(y + 5) + 1(z + 4) - 8 = 0$$

$$\therefore -x + 3y + 10z - 58 = 0$$

$$\therefore x - 3y - 10z + 58 = 0.$$

∴ Equations of required lines are
$3x - 2y + 4z + 3 = 0$ and $x - 3y - 10z + 58 = 0$.
Let l, m, n be the direction ratios of the line
∴ $3l - 2m + 4n = 0$ and $l - 3m - 10n = 0$.
∴ By Cramer's rule;

$$\frac{l}{\begin{vmatrix} -2 & 4 \\ -3 & -10 \end{vmatrix}} = \frac{m}{\begin{vmatrix} 4 & -10 \\ 3 & 1 \end{vmatrix}} = \frac{n}{\begin{vmatrix} 3 & -2 \\ 1 & -3 \end{vmatrix}}$$

$$\therefore \frac{l}{32} = \frac{m}{34} = \frac{n}{-7}.$$

∴ The equation of the tangent line to the given circle at $(-3, 5, 4)$ is

$$\frac{x + 3}{32} = \frac{y - 5}{34} = \frac{z - 4}{-7}.$$

32) Show that the plane $2x - 2y - z + 12 = 0$ touches the sphere $x^2 + y^2 + z^2 - 2x - 4y + 2z = 3$ and finds the point of contact.

Sol. The equation of sphere is

$$x^2 + y^2 + z^2 - 2x - 4y + 2z - 3 = 0.$$

∴ Comparing with the general equation, we get

$$2u = -2;\ 2v = -4;\ 2w = 2;\ d = -3$$

$$\therefore u = -1;\ v = -2;\ w = 1.$$

∴ The center of the sphere is $C(1, 2, -1)$ and the radius of the sphere $= \sqrt{1 + 4 + 1 + 3} = 3$. Length of the perpendicular from center to plane $= \frac{2(1) - 2(2) + 1(-1) + 12}{\sqrt{4 + 4 + 1}} = 3.$

Since the length of the perpendicular from the center of the sphere to the plane = radius of the sphere.
∴ The plane touches the sphere.
∴ Direction ratios of the radius through the point of contact P of the given tangent plane are $(2, -2, 1)$.
∴ the Equation of CP is $\frac{x-1}{2} = \frac{y-2}{-2} = \frac{z+1}{1} = r$.
∴ Coordinates of P are $(2r + 1, -2r + 2, r - 1)$.
The point lies on the plane and satisfies the equation of the plane

$$2x - 2y + z + 12 = 0$$

∴ $2(2r + 1) - 2(-2r + 2) + (r - 1) + 12 = 0$
∴ $4r + 2 + 4r - 4 - r - 1 + 12 = 0$
∴ $9r + 9 = 0$
∴ $r = -1$.

∴ The point of contact is $(-1, 4, 2)$.

33) Find the coordinates of the points of the sphere $x^2 + y^2 + z^2 - 4x + 2y = 4$ the tangent planes which are parallel to the plane $2x - y + 2z = 1$.

Sol. The equation of sphere is

$$x^2 + y^2 + z^2 - 4x + 2y - 4 = 0$$

∴ Comparing with the general equation

$$2u = -4; 2v = 2; 2w = 0; d = -4$$

∴ $u = -2; v = 1; w = 0$.

∴ The center of the sphere is $C(2, -1, 0)$ and the radius of the sphere is $\sqrt{4 + 1 + 0 + 4} = 3$
The equation of the plane is $2x - y + 2z - 1 = 0$
∴ Equation of tangent plane parallel to the given plane is

$$2x - y + 2z + k = 0.$$

Given plane is tangent if the perpendicular distance from the center of the sphere to the plane = radius of the sphere

$$\therefore 3 = \frac{2(2) + (-1)(-1) + 2(0) + k}{\sqrt{4 + 1 + 4}}$$

$$\therefore \pm 9 = 5 + k$$
$$\therefore 9 = 5 + k \text{ or } -9 = 5 + k$$
$$\therefore k = 4 \text{ or } k = -14.$$

∴ The equations of two tangent planes to the given plane is
$2x - y + 2z + 4 = 0$ and $2x - y + 2z - 14 = 0$.
∴ The equation of radius perpendicular to these planes are

$$\frac{x-2}{2} = \frac{y+1}{-1} = \frac{z}{2} = r_1.$$

Let P be the point of contact of plane $2x - y + 2z + 4 = 0$.
∴ Coordinates of P are $(2r_1 + 2; -r_1 - 1; 2r_1)$

$$\therefore 2(2r_1 + 2) - (-r_1 - 1) + 2(2r_1) + 4 = 0$$
$$\therefore 4r_1 + 4 + r_1 + 1 + 4r_1 + 4 = 0$$
$$\therefore 9r_1 + 9 = 0$$
$$\therefore r_1 = -1$$

∴ Coordinates are $(0, 0, -2)$.
Let Q be the point of contact of plane $2x - y - 2z - 14 = 0$.
∴ Coordinates of Q are $(2r_2 + 2; -r_2 - 1; 2r_2)$

$$\therefore 2(2r_2 + 2) - (-r_2 - 1) + 2(2r_2) + 4 = 0$$
$$\therefore 4r_2 + 4 + r_2 + 1 + 4r_2 - 14 = 0$$
$$\therefore 9r_2 = 9$$
$$\therefore r_2 = 1$$

∴ Coordinates are $(4, -2, 2)$.

34) Obtain the equation of the tangent planes to the sphere $x^2 + y^2 + z^2 + 6x - 2z + 1 = 0$ which passes through line $3(16 - x) = 3z = 2y + 30$.

Sol. The equation of sphere is $x^2 + y^2 + z^2 + 6x - 2z + 1 = 0$.
∴ The center of the sphere C is (-3, 0, 1) and the radius of the sphere is $\sqrt{(-3)^2 + 0^2 + (1)^2 - 1} = 3$.
The equation of the line is $3(16 - x) = 2y + 30 = 3z$.

The intersection of lines is a plane $3(16 - x) = 3z$ and $3z = 2y + 30$.

$$\therefore 16 - x = z \quad \text{and} \quad 2y - 3z + 30 = 0.$$

Any plane passing through this line is

$$(x + z - 16) + \lambda(2y - 3z + 30) = 0$$
$$\therefore x + 2\lambda y + z(1 - 3\lambda) + (30\lambda - 16) = 0. \tag{3.56}$$

Plane (3.56) will touch the given sphere if the length of the perpendicular from (-3, 0, 1) to plane (3.56) radius of the sphere

$$\therefore 3 = \frac{(1)(-3) + (0)(2\lambda) + (1)(1 - 3\lambda) + 30\lambda - 16}{\sqrt{1 + 4\lambda^2 + (1 - 3\lambda)^2}}$$

$$\therefore 3 = \frac{27\lambda - 18}{\sqrt{13\lambda^2 - 6\lambda + 2}}$$

$$\therefore 9(13\lambda^2 - 6\lambda + 2) = [9(3\lambda - 2)]^2$$

$$\therefore 13\lambda^2 - 6\lambda + 2 = 81\lambda^2 - 108 + 36$$

$$\therefore 2\lambda^2 - 3\lambda + 1 = 0$$

$$\therefore \lambda = 1; \lambda = \frac{1}{2}.$$

Substituting the value of λ in Equation (3.56), we get

$$x + 2(1)y + z(1 - 3) + (30 - 16) = 0$$

$$\therefore x + 2y - 2z + 14 = 0.$$

When $\lambda = \frac{1}{2}$

$$x + 2\left(\frac{1}{2}\right)y + z\left(1 - \frac{3}{2}\right) + \left(\frac{30}{2} - 16\right) = 0$$

$$2x + 2y - z - 2 = 0.$$

35) Obtain the equations of the sphere which passes through the circle $x^2 + y^2 + z^2 - 2x + 2y + 4z - 3 = 0$; $2x + y + z = 4$ and touch the plane $3x + 4y = 14$.

3.10 Equation of the Normal to the Sphere

Sol. The equation of the sphere is $S + \lambda P = 0$.

$\therefore (x^2 + y^2 + z^2 - 2x + 2y + 4z - 3) + \lambda(2x + y + z - 4) = 0$

$\therefore x^2 + y^2 + z^2 + (-2 + 2\lambda)x + (2 + \lambda)y + (4 + \lambda)z$
$\qquad - (3 + 4\lambda) = 0.$ (3.57)

$\therefore$ Center is $\left[1 - \lambda; \left(\frac{2+\lambda}{2}\right); \left(\frac{-4-\lambda}{2}\right)\right]$.

$\therefore$ Radius is $\sqrt{(-1 + \lambda^2) + \left(\frac{2+\lambda}{2}\right)^2 + \left(\frac{4+\lambda}{2}\right)^2 + (3 + 4\lambda)}$

$\qquad = \dfrac{1}{2}\sqrt{6\lambda^2 + 20\lambda + 36}.$

Length of the perpendicular from center to the plane

$3x + 4y - 14 = \dfrac{3(1-\lambda) + 4\left(\frac{-2-\lambda}{2}\right) - 14}{\sqrt{3^2 + 4^2}}$

$\therefore \dfrac{3(1-\lambda) + 4\left(\frac{-2-\lambda}{2}\right) - 14}{5} = \dfrac{1}{2}\sqrt{6\lambda^2 + 20\lambda + 36}$

$\therefore \dfrac{-15 - 5\lambda}{5} = \dfrac{1}{2}\sqrt{6\lambda^2 + 20\lambda + 36}$

$\therefore -2(\lambda + 3) = \sqrt{6\lambda^2 + 20\lambda + 36}$

$\therefore 4(\lambda + 3)^2 = 6\lambda^2 + 20\lambda + 36$

$\therefore 2\lambda^2 - 4\lambda = 0$

$\therefore \lambda = 0 \text{ or } \lambda = 2.$

$\therefore$ Substituting the value of λ in Equation (3.57), we get

$\therefore x^2 + y^2 + z^2 - 2x + 2y - 4z - 3 = 0$

and when $\lambda = 2$ we get, $x^2 + y^2 + z^2 + 2x + 4y + 6z - 11 = 0$.

36) Show that the sphere $x^2 + y^2 + z^2 = 25$; $x^2 + y^2 + z^2 + 24x - 40y - 18z + 225 = 0$ touch externally and find the point of the contact.

Sol. The given spheres are

$$x^2 + y^2 + z^2 = 25,$$ (3.58)

126 Sphere

$$x^2 + y^2 + z^2 - 24x - 40y - 18z + 225 = 0. \qquad (3.59)$$

Suppose two spheres touch each other at the point (α, β, γ).

$\therefore$ The equations of tangent planes to these spheres are

$$\alpha x + \beta y + \gamma z = 25, \qquad (3.60)$$

and

$$x(\alpha - 12) + y(\beta - 20) + z(\gamma - 9) - 12\alpha - 20\beta - 9\gamma + 225 = 0$$
$$\therefore x(\alpha - 12) + y(\beta - 20) + z(\gamma - 9) = 12\alpha + 20\beta + 9\gamma - 225 \qquad (3.61)$$

If two spheres touch each other than these two tangent planes must coincide.

$\therefore$ Comparing the coefficient of (3.60) and (3.61), we get

$$\frac{\alpha - 12}{\alpha} = \frac{\beta - 20}{\beta} = \frac{\gamma - 9}{\gamma} = \frac{12\alpha + 20\beta + 9\gamma - 225}{25} = k\alpha - 12$$

$$= k\alpha; \beta - 20 = k\beta; \gamma - 9 = k\gamma; 12\alpha + 20\beta + 9\gamma - 225 = 25k$$

$$\therefore \alpha - k\alpha = 12\beta - k\beta = 20\gamma - k\gamma = 9$$

$$\therefore \alpha(1-k) = 12\beta(1-k) = 20\gamma(1-k) = 9$$

$$\therefore \alpha = \frac{12}{1-k}\beta = \frac{20}{1-k}\lambda = \frac{9}{1-k}$$

$$\therefore 12\alpha + 20\beta + 9\gamma - 225 = 25k$$

$$\therefore 12\left(\frac{12}{1-k}\right) + 20\left(\frac{20}{1-k}\right) + 9\left(\frac{9}{1-k}\right) - 225 = 25k$$

$$\therefore 144 + 400 + 81 - 225(1-k) = 25k(1-k)$$

$$\therefore k^2 + 8k + 16 = 0$$

$$\therefore (k+4)^2 = 0$$

$$\therefore k = -4.$$

$\therefore$ The point of contact is $(\alpha, \beta, \gamma) = \left(\frac{12}{5}; \frac{20}{5}; \frac{9}{5}\right)$.

$\therefore$ The radius of the spheres is 5 and $\sqrt{12^2 + 20^2 + 9^2 - 225} = 20$.

$\therefore$ Sum of radii $= 5 + 20 = 25$.

Center are $(0, 0, 0)$ and $(12, 20, 9)$

$\therefore$ Distance between centers $= \sqrt{12^2 + 20^2 + 9^2} = 25$.

$\therefore$ Distance between centers = sum of radii

$\therefore$ Spheres touch each other externally.

3.11 Orthogonal Sphere

Two spheres are said to cut each other orthogonally if the tangent planes at a point of contact are perpendicular to each other.

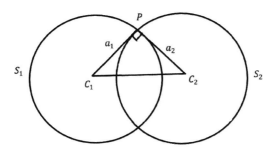

The condition for the spheres to cut orthogonally is

$$2u_1u_2 + 2v_1v_2 + 2w_1w_2 = d_1 + d_2.$$

36) Find the equation of the sphere that passes through the circle $x^2 + y^2 + z^2 - 2x + 3y - 4z + 6 = 0; 3x - 4y + 5z - 15 = 0$ and cuts the sphere $x^2 + y^2 + z^2 + 2x + 4y - 6z + 11 = 0$ orthogonally.

Sol. Given the equation of the circle is

$$x^2 + y^2 + z^2 - 2x + 3y - 4z + 6 = 0 \qquad (3.62)$$

$$3x - 4y + 5z - 15 = 0.$$

Equation of sphere through this circle is

$$x^2 + y^2 + z^2 - 2x + 3y - 4z + 6 + \lambda(3x - 4y + 5z - 15) = 0 \quad (3.63)$$

$$\therefore x^2 + y^2 + z^2 + (3\lambda - 2)x + (3 - 4\lambda)y + (5\lambda - 4)z + 6 - 15\lambda = 0$$

$$\therefore \text{Center is } \left[-\left(\frac{3\lambda - 2}{2}\right); -\left(\frac{3 - 4\lambda}{2}\right); -\left(\frac{5\lambda - 4}{2}\right) \right].$$

Another equation of sphere is

$$x^2 + y^2 + z^2 + 2x + 4y - 6z + 11 = 0.$$

$\therefore$ Center is $(-1, -2, 3)$.

128 Sphere

The condition of the orthogonal intersection of spheres is

$$2uu_1 + 2vv_1 + 2ww_1 = d + d_1$$

$$\therefore 2\left(\frac{3\lambda - 2}{2}\right)(1) + 2\left(\frac{3 - 4\lambda}{2}\right)(2) + 2\left(\frac{5\lambda - 4}{2}\right)(-3) = 6 - 15\lambda + 11$$

$$\therefore 3\lambda - 2 + 6 - 8\lambda - 15\lambda + 12 = 17 - 15\lambda$$

$$\therefore -5\lambda = 1$$

$$\therefore \lambda = -\frac{1}{5}.$$

$\therefore$ Substituting the value of λ in equation (3.63), we get

$$x^2 + y^2 + z^2 + \left(\frac{-3}{5} - 2\right)x + \left(3 + \frac{4}{5}\right)y + (-1 - 4)z + 6 + \frac{15}{5} = 0$$

$$\therefore x^2 + y^2 + z^2 - \frac{13}{5}x + \frac{19}{5}y - 5z + \frac{45}{5} = 0$$

$$\therefore 5x^2 + 5y^2 + 5z^2 - 13x + 19y - 25z + 45 = 0.$$

37) Find the equation of the sphere that passes through the two points (0, 3, 0); (-2, -1, -4) and cuts orthogonally the two spheres $x^2 + y^2 + z^2 + x - 3z - 2 = 0$ and $2\left(x^2 + y^2 + z^2\right) + x + 3y + 4 = 0$.

Sol. The general equation of sphere is

$$x^2 + y^2 + z^2 + 2ux + 2vy + 2wz + d = 0. \quad (3.64)$$

It passes through the point (0, 3, 0)

$$\therefore 0 + 9 + 0 + 0 + 6v + 0 + d = 0$$

$$\therefore 6v + d = -9. \quad (3.65)$$

It passes through point (-2, -1, -4)

$$\therefore 4 + 1 + 16 - 4u - 2v - 8w + d = 0$$

$$\therefore -4u - 2v - 8w + d = -21. \quad (3.66)$$

Sphere (3.64) also cuts the sphere
$x^2 + y^2 + z^2 + x - 3z - 2 = 0$ orthogonally.

$$\therefore 2u\left(\frac{1}{2}\right) + 2v\left(0\right) + 2w\left(\frac{-3}{2}\right) = d + (-2)$$

$$\therefore u - 3w - d = -2. \quad (3.67)$$

Sphere (3.64) also cuts the sphere
$$2(x^2 + y^2 + z^2) + x + 3y + 4 = 0$$
$$\therefore x^2 + y^2 + z^2 + \tfrac{x}{2} + \tfrac{3y}{2} + 2 = 0 \text{ orthogonally.}$$
$$\therefore 2u\left(\tfrac{1}{4}\right) + 2v\left(\tfrac{3}{4}\right) + 2w(0) = d + 2 \qquad (3.68)$$
$$\therefore u + 3v - 2d = 4.$$

From (3.65) $\Rightarrow v = \tfrac{-9-d}{6}$
From (3.68) $\Rightarrow u = 4 + 2d - 3v = 4 + 2d - 3\left(\tfrac{-9-d}{6}\right)$
$$\therefore u = \tfrac{17 + 5d}{2}$$

From (3.67) $\Rightarrow 3w = u - d + 2 = \tfrac{17+5d}{2} - d + 2$
$$\therefore w = \tfrac{21 + 3d}{6}$$

From (3.66) $\Rightarrow -4\left(\tfrac{17+5d}{2}\right) - 2\left(\tfrac{-9-d}{6}\right) - 8\left(\tfrac{21+3d}{6}\right) + d = -21$
$$\therefore -2(17 + 5d) + \left(\tfrac{9+d}{3}\right) - 4\left(\tfrac{21+3d}{3}\right) + d = -21$$
$$\therefore -102 - 30d + 9 + d - 84 - 12d + 3d = -63$$
$$\therefore -38d = 144$$
$$\therefore d = -3,\ u = 1,\ v = -1,\ w = 2$$

The equation of sphere is $x^2 + y^2 + z^2 + 2x - 2y - 4z - 3 = 0$.

Exercise:

1) Find the equation of the sphere whose center is $(2, -3, 4)$ and radius 5.

Answer: $x^2 + y^2 + z^2 - 4x + 6y - 8z + 4 = 0$

2) Find the center and radius of the following sphere
2.1 $\quad 2x^2 + 2y^2 + 2z^2 - 2x + 4y - 2z - 3 = 0$
2.2 $\quad x^2 + y^2 + z^2 - 2x + 4y - 6z = 2.$

Answer: (i) $\left(\tfrac{1}{2}, -1, \tfrac{-1}{2}\right), 0$ (ii) $(1, -2, 3), 4$

3) Prove that the equation $ax^2 + ay^2 + az^2 + 2ux + 2vy + 2wz + d = 0$ represents a sphere. Find its radius and center.

Answer: $\frac{1}{a}\sqrt{u^2 + v^2 + w^2 - ad}$, $\left(\frac{-u}{a}, \frac{-v}{a}, \frac{-w}{a}\right)$

4) Find the equation of the sphere through the points.

4.1 $(4, -1, 2), (0, -2, 3), (1, 5, -1), (2, 0, 1)$

4.2 $(0, 0, 0), (0, 1, -1), (-1, 2, 0), (1, 2, 3)$.

Answer: (i) $x^2 + y^2 + z^2 - 4x - 14y - 22z + 25 = 0$

(ii) $7(x^2 + y^2 + z^2) - 15x - 25y - 11z = 0$

5) Find the equation of a sphere that passes through $(0, 0, 0)$ and which has its center at $\left(\frac{1}{2}, \frac{1}{2}, 0\right)$.

Answer: $2x^2 + 2y^2 + 2z^2 = 1$

6) Find the equation of the sphere circumscribing the tetrahedron whose faces are $\frac{y}{b} + \frac{z}{c} = 0$, $\frac{z}{c} + \frac{x}{a} = 0$, $\frac{x}{a} + \frac{y}{b} = 0$, $\frac{x}{a} + \frac{y}{b} + \frac{z}{c} = 1$.

Answer: $\frac{x^2+y^2+z^2}{a^2+b^2+c^2} - \frac{x}{a} - \frac{y}{b} - \frac{z}{c} = 0$

7) Find the equation of the sphere which passes through the origin and makes equal intercepts of unit length on the axes.

Answer: $x^2 + y^2 + z^2 - x - y - z = 0$

8) A plane passes through a fixed point (a, b, c) and cuts the axes in A, B, C. Show that the locus of the center of the sphere $OABC$ is $\frac{a}{x} + \frac{b}{y} + \frac{c}{z} = 2$.

9) A sphere of constant radius k passes through the origin and meets the axes in A, B, C. Prove that the centroid of the triangle ABC lies on the sphere.

10) Find the equation of the sphere on the join of $(2, -3, 1)$ and $(1, -2, -1)$ as diameter.

Answer: $x^2 + y^2 + z^2 - 3x + 5y + 7 = 0$

11) Find the equation of the sphere on the join of $(1, -2, 3)$ and $(2, 1, 0)$ as diameter.

Answer: $x^2 + y^2 + z^2 - 3x - y - 3z = 0$

Exercise: 131

12) Find the center and radius of the circle given by the equation $x^2 + y^2 + z^2 - 6x - 4y + 12z - 36 = 0$ and $x + 2y - 2z = 1$.

Answer: $(1, -2, -2)$, 7

13) Find the center and radius of the circle in which the sphere $x^2 + y^2 + z^2 + 2x - 2y - 4z - 19 = 0$ is cut by the plane $x + 2y + 2z + 7 = 0$.

Answer: $\left(\frac{-7}{3}, \frac{-5}{3}, \frac{-2}{3}\right)$, 3

14) Show that the radius of the circle
$$x^2 + y^2 + z^2 + x + y + z - 4 = 0, \ x + y + z = 0 \text{ is } 2.$$

15) Find the center and radius of the circle
$$x^2 + y^2 + z^2 - 8x + 4y + 8z - 45 = 0, \ x - 2y + 3z = 3.$$

Answer: $\left(\frac{9}{2}, -3, \frac{-5}{2}\right)$, $\sqrt{\frac{155}{2}}$

16) Find the equation of that plane that cuts the sphere $x^2 + y^2 + z^2 = a^2$ in a circle whose center is (α, β, γ).

Answer: $(x - \alpha) + \beta(y - \beta) + \gamma(z - \gamma) = 0$

17) A circle with center $(2, 3, 0)$ and radius 1 is drawn in the plane $z = 0$. Find the equation of the sphere which passes through this circle and the point $(1, 1, 1)$.

Answer: $x^2 + y^2 + z^2 - 4x - 6y - 5z + 12 = 0$

18) Prove that the plane $x + 2y - z = 4$ cuts the sphere $x^2 + y^2 + z^2 - x + z - 2 = 0$ in a circle of radius unity, and find the equation of the sphere which has this circle for one of the great circles.

19) Find the sphere for which the circle $x^2 + y^2 + z^2 + 10y - 4z - 8 = 0$, $x + y + z = 3$ is a great circle.

Answer: $x^2 + y^2 + z^2 - 4x + 6y - 8z + 4 = 0$

20) Prove that the circles

$$x^2 + y^2 + z^2 - 2x + 3y + 4z - 5 = 0, 5y + 6z + 1 = 0;$$
$$x^2 + y^2 + z^2 - 3x - 4y + 5z - 6 = 0, x + 2y - 7z = 0$$

lie on the same sphere and find its equation.

21) Find the equations of the tangent line in the symmetrical form to circle $3x^2 + 3y^2 + 3z^2 - 2x - 3y - 4z - 22 = 0, 3x + 4y + 5z - 26 = 0$ at the point $(1, 2, 3)$.

Answer: $\frac{x-1}{1} = \frac{y-2}{-2} = \frac{z-3}{1}$

22) Show that $2x - y - 2z = 4$ is the tangent plane to the sphere $x^2 + y^2 + z^2 + 2x - 6y + 1 = 0$. Also, find the point of contact.

Answer: $(1, 2, -2)$

23) Show that the spheres $x^2 + y^2 + z^2 = 64$ and $x^2 + y^2 + z^2 - 12x + 4y - 6z + 48 = 0$ touch internally and find their point of contact.

Answer: $\left(\frac{48}{7}, \frac{-16}{7}, \frac{24}{7}\right)$

24) Show that the spheres $x^2 + y^2 + z^2 + 6y + 2z + 8 = 0$ and $x^2 + y^2 + z^2 + 6x + 8y + 4z + 20 = 0$ are orthogonal. Find their plane of intersection.

Answer: $3x + y + z + 6 = 0$

4

Cone

4.1 Definition

A cone is a surface generated by a straight line that passes through a fixed point and satisfies one more condition.

i.e., it intersects a given curve or touches a given surface.

The fixed point is called the vertex and the given curve is called the guiding curve of the cone.

The individual straight line on the surface of a cone is called its generator.

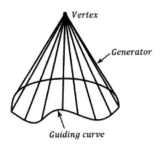
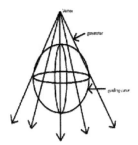

4.2 Equation of a Cone with a Conic as Guiding Curve

To find the equation of a cone with vertex (α, β, γ) and whose generators intersect the conic $ax^2 + by^2 + 2hxy + 2gx + 2fy + c = 0$ and $z = 0$.

Consider a straight line passing through a fixed point (α, β, γ) and having direction ratios (l, m, n).

∴ Equation of line will be

$$\frac{x-\alpha}{l} = \frac{y-\beta}{m} = \frac{z-\gamma}{n}. \tag{4.1}$$

This line will be a generator of the cone if and only if it intersects the given curve.

Moreover, line (4.1) touches the plane $z = 0$.

i.e., $\frac{x-\alpha}{l} = \frac{y-\beta}{m} = \frac{-\gamma}{n}$

$\therefore \frac{x-\alpha}{l} = \frac{-\gamma}{n}$ and $\frac{y-\beta}{m} = \frac{-\gamma}{n}$

$\therefore x = \alpha - \frac{l\gamma}{n}$ and $y = \beta - \frac{m\gamma}{n}$ and $z = 0$

$\therefore \left(\alpha - \frac{l\gamma}{n}; \beta - \frac{m\gamma}{n}; 0\right)$ is the point which lies on the given conic.

$\therefore$ It satisfies the given equation of conic

$$ax^2 + by^2 + 2hxy + 2gx + 2fy + c = 0$$

$$\therefore a\left(\alpha - \frac{l\gamma}{n}\right)^2 + b\left(\beta - \frac{m\gamma}{n}\right)^2 + 2h\left(\alpha - \frac{l\gamma}{n}\right)\left(\beta - \frac{m\gamma}{n}\right) + 2g\left(\alpha - \frac{l\gamma}{n}\right)$$

$$+ 2f\left(\beta - \frac{m\gamma}{n}\right) + c = 0$$

$$\therefore a(\alpha n - l\gamma)^2 + b(\beta n - m\gamma)^2 + 2h(\alpha n - l\gamma)(\beta n - m\gamma)$$
$$+ 2gn(\alpha n - l\gamma) + 2fn(\beta n - m\gamma) + cn^2 = 0. \quad (4.2)$$

This is the condition for the line (4.1) to intersect the given conic.

Eliminating $l, m,$ and n from Equation (4.2), we get the equation of cone i.e., replace l, m and n by $x - \alpha, y - \beta$ and $z - \gamma$ in Equation (4.2)

$$\therefore a[\alpha(z-\gamma) - \gamma(x-\alpha)]^2 + b[\beta(z-\gamma) - (y-\beta)\gamma]^2$$
$$+ 2h[\alpha(z-\gamma) - \gamma(x-\alpha)][\beta(z-\gamma) - \gamma(y-\beta)]$$
$$+ 2g(z-\gamma)[\alpha(z-\gamma) - \gamma(x-\alpha)]$$
$$+ 2f(z-\gamma)[\beta(z-\gamma) - \gamma(y-\beta)] + c(z-\gamma)^2 = 0$$
$$\therefore a(\alpha z - x\gamma)^2 + b(\beta z - y\gamma)^2 + 2h(\alpha z - x\gamma)(\beta z - y\gamma)$$
$$+ 2g(z-\gamma)(\alpha z - x\gamma) + 2f(\beta z - y\gamma)(z-\gamma) + c(z-\gamma)^2 = 0,$$

which is the required equation of the cone.

1) Find the equation of the cone whose generators pass through the point (α, β, γ) and have their direction cosines satisfying the relation $al^2 + bm^2 + cn^2 = 0$.

Sol. Given all the generators pass through (α, β, γ).

4.2 Equation of a Cone with a Conic as Guiding Curve

∴ (α, β, γ) is the vertex of the cone.
∴ Equation of line passing through (α, β, γ) is

$$\frac{x-\alpha}{l} = \frac{y-\beta}{m} = \frac{z-\gamma}{n} = k$$

∴ $l = \dfrac{x-\alpha}{k}; m = \dfrac{y-\beta}{k}; n = \dfrac{z-\gamma}{k},$

where l, m and n satisfy the relation $al^2 + bm^2 + cn^2 = 0$

$$\therefore a\left[\frac{x-\alpha}{k}\right]^2 + b\left[\frac{y-\beta}{k}\right]^2 + c\left[\frac{z-\gamma}{k}\right]^2 = 0$$

$$\therefore a(x-\alpha)^2 + b(y-\beta)^2 + c(z-\gamma)^2 = 0$$

is the required equation of the cone.

2) Find the equation of the cone whose vertex is the point $(1,1,0)$ and whose guiding curve is $y = 0; x^2 + z^2 = 4$.

Sol. Equation of the generator through the vertex $(1,1,0)$ is

$$\frac{x-1}{l} = \frac{y-1}{m} = \frac{z-0}{n} = k. \tag{4.3}$$

∴ Any point on this generator is $P(lk+1; mk+1; nk)$.
Suppose the generator meets the guiding curve

$$y = 0 \text{ and } x^2 + y^2 = 4 \tag{4.4}$$

∴ $mk + 1 = 0$ and $(lk+1)^2 + n^2 k^2 = 4$

∴ $k = \dfrac{-1}{m}$

∴ $\left(\dfrac{-l}{m} + 1\right)^2 + \dfrac{n^2}{m^2} = 4$

∴ $(m-l)^2 + n^2 = 4m^2$.

Substituting the value of l, m and n, we get

$$\therefore \left[\left(\frac{y-1}{k}\right) - \left(\frac{x-1}{k}\right)\right]^2 + \frac{z^2}{k^2} = 4\left(\frac{y-1}{k}\right)^2$$

∴ $(y-x)^2 + z^2 = 4(y-1)^2$

∴ $x^2 + y^2 + z^2 - 2xy = 4y^2 - 8y + 4$

∴ $x^2 - 3y^2 + z^2 - 2xy + 8y - 4 = 0$ is the required equation of the cone.

3) Find the equation of the cone whose vertex is (α, β, γ) and whose base is $y^2 = 4ax, z = 0$.

Sol. The given equation of base conic is

$$y^2 = 4ax, \quad z = 0. \tag{4.5}$$

Equation of any line through (α, β, γ) are

$$\frac{x - \alpha}{l} = \frac{y - \beta}{m} = \frac{z - \gamma}{n}. \tag{4.6}$$

This meets the plane $z = 0$,

$$\frac{x - \alpha}{l} = \frac{y - \beta}{m} = \frac{0 - \gamma}{n}$$

$$\therefore x = \alpha - \frac{l\gamma}{n}, y = \beta - \frac{m\gamma}{n}.$$

Substituting the values of x and y in a given base conic, we get

$$\left(\beta - \frac{m\gamma}{n}\right)^2 = 4a\left(\alpha - \frac{l\gamma}{n}\right)$$

$$\therefore \left(\frac{\beta n - m\gamma}{n}\right)^2 = 4a\left(\frac{\alpha n - l\gamma}{n}\right). \tag{4.7}$$

Eliminating $l, m,$ and n from Equations (4.6) and (4.7), we get

$$[\beta(z - \gamma) - \gamma(y - \beta)]^2 = 4a(z - \gamma)[-(x - \alpha)\gamma + \alpha(z - \gamma)]$$
$$\therefore [\beta z - \beta\gamma - y\gamma + \gamma\beta]^2 = 4a(z - \gamma)[-x\gamma + \alpha\gamma + \alpha z - \alpha\gamma]$$
$$\therefore [\beta z - y\gamma]^2 = 4a(z - \gamma)(\alpha z - x\gamma)$$

which is the required equation of the cone.

4) Show that the equation of the cone whose vertex is the origin and whose base is the circle through the three points $(a, 0, 0), (0, b, 0), (0, 0, c)$ is $\Sigma a \left(b^2 + c^2\right) yz = 0$.

Sol. Let the given points be $A(a, 0, 0); B(0, b, 0)$ and $C(0, 0, c)$
∴ The equation of the circle ABC is

$$x^2 + y^2 + z^2 - ax - by - cz = 0 \text{ and } \frac{x}{a} + \frac{y}{b} + \frac{z}{c} = 1. \tag{4.8}$$

4.2 Equation of a Cone with a Conic as Guiding Curve

Let $P(x_1, y_1, z_1)$ be any point on the surface of the cone whose vertex is the origin.

∴ The equation of generator OP is

$$\frac{x-0}{x_1-0} = \frac{y-0}{y_1-0} = \frac{z-0}{z_1-0} = k$$

$$\therefore \frac{x}{x_1} = \frac{y}{y_1} = \frac{z}{z_1} = k. \tag{4.9}$$

∴ Any point on the generator is $Q(kx_1, ky_1, kz_1)$.

Suppose the generator (4.9) meets the circle (4.8) at this point

$$\therefore k^2 x_1^2 + k^2 y_1^2 + k_1^2 z_1^2 - akx_1 - bxy_1 - ckz_1 = 0$$

and $\frac{kx_1}{a} + \frac{ky_1}{b} + \frac{kz_1}{c} = 1$

$$\therefore k^2 \left(x_1^2 + y_1^2 + z_1^2 \right) = k(ax_1 + by_1 + cz_1)$$

$$\therefore k \left(x_1^2 + y_1^2 + z_1^2 \right) = ax_1 + by_1 + cz_1$$

and

$$k \left(\frac{x_1}{a} + \frac{y_1}{b} + \frac{z_1}{c} \right) = 1.$$

Eliminating k, we get

$$x_1^2 + y_1^2 + z_1^2 = (ax_1 + by_1 + cz_1) \left(\frac{x_1}{a} + \frac{y_1}{b} + \frac{z_1}{c} \right)$$

$$\therefore x_1^2 + y_1^2 + z_1^2 = x_1^2 + \frac{ax_1 y_1}{b} + \frac{ax_1 z_1}{c} + \frac{bx_1 y_1}{a} + y_1^2 + \frac{by_1 z_1}{c}$$

$$+ \frac{cz_1 x_1}{a} + \frac{cz_1 y_1}{b} + z_1^2$$

$$\therefore x_1 y_1 \left(\frac{a}{b} + \frac{b}{a} \right) + y_1 z_1 \left(\frac{b}{c} + \frac{c}{b} \right) + x_1 z_1 \left(\frac{c}{a} + \frac{a}{c} \right) = 0$$

$$\therefore cx_1 y_1 \left(a^2 + b^2 \right) + ay_1 z_1 \left(b^2 + c^2 \right) + bx_1 z_1 \left(c^2 + a^2 \right) = 0.$$

∴ Generalizing x_1, y_1, z_1, we get

$$c \left(a^2 + b^2 \right) xy + a \left(b^2 + c^2 \right) yz + b \left(c^2 + a^2 \right) xz = 0$$

$$\therefore \sum a \left(b^2 + c^2 \right) yz = 0$$

which is the required equation of the cone.,

5) Find the equation of the cone whose vertex is the point $(1, 2, 3)$ and guiding curve the circle $x^2 + y^2 + z^2 = 4$; $x + y + z = 1$.

Sol. Equation of any generator through the vertex $P(1, 2, 3)$ is

$$\frac{x-1}{l} = \frac{y-2}{m} = \frac{z-3}{n} = \frac{x+y+z-6}{l+m+n}. \tag{4.10}$$

If it meets the given conic and plane $x + y + z = 1$

∴ Equation (4.10) becomes

$$\frac{x-1}{l} = \frac{y-2}{m} = \frac{z-3}{n} = \frac{-5}{l+m+n}$$

$$\therefore x = 1 - \frac{5l}{l+m+n}; y = 2 - \frac{5m}{l+m+n}; z = 3 - \frac{5n}{l+m+n}$$

∴ Generator (4.10) meets the given plane at the point

$$\left(\frac{m+n-4l}{l+m+n}; \frac{2l+2n-3m}{l+m+n}; \frac{3l+3m-2n}{l+m+n}\right).$$

If this point lies on the surface $x^2 + y^2 + z^2 = 4$

$$\therefore (m+n-4l)^2 + (2l+2n-3m)^2 + (3l+3m-2n)^2$$
$$= 4(l+m+n)^2.$$

Eliminating l, m, n from Equation (4.10), we get

$$\therefore [(y-2) + (z-3) - 4(x-1)]^2$$
$$+ [2(x-1) + 2(z-3) - 3(y-2)]^2$$
$$+ [3(x-1) + 3(y-2) - 2(z-3)]^2$$
$$= 4[(x-1) + (y-2) + (z-3)]^2$$
$$\therefore (-4x + y + z - 1)^2 + (2x - 3y + 2z - 2)^2$$
$$+ (3x + 3y - 2z - 3)^2 = 4(x + y + z - 6)^2$$
$$\therefore 5x^2 + 3y^2 + z^2 - 6yz - 4zx - xy + 6x + 8y + 10z - 26 = 0.$$

4.3 Enveloping Cone to a Surface

Enveloping cone to a surface is a cone formed by the tangent lines to a surface drawn from a given point.

Enveloping Cone to a Sphere

Enveloping cone to a sphere is a surface generated by a straight line passing through a fixed point and touching a given sphere.

4.3 Enveloping Cone to a Surface

Equation of an enveloping cone to a sphere:

To find the equation of on enveloping cone to a sphere $x^2 + y^2 + z^2 = a^2$ with vertex (α, β, γ).

Let (l, m, n) be the direction ratios of the straight line passing through the vertex (α, β, γ).

$\therefore$ Equation of straight line passing through (α, β, γ) is

$$\frac{x-\alpha}{l} = \frac{y-\beta}{m} = \frac{z-\gamma}{n} = t \tag{4.11}$$

$$\therefore x = \alpha + lt; y = \beta + mt; z = \gamma + nt. \tag{4.12}$$

The line (4.11) will be the generator of the given curve if and only if it touches the given sphere.

$\therefore$ The point of intersection of the line (4.11) and the sphere is

$$x^2 + y^2 + z^2 = a^2$$

$$\therefore (\alpha + lt)^2 + (\beta + mt)^2 + (\gamma + nt)^2 = a^2$$

$$\therefore t^2\left(l^2 + m^2 + n^2\right) + 2t\left(l\alpha + \beta m + \gamma n\right) + \left(\alpha^2 + \beta^2 + \gamma^2 - a^2\right) = 0 \tag{4.13}$$

which is a quadratic equation in t; so, the line will touch the sphere if the roots of the quadratic equation are equation.

$$\therefore \Delta = 0$$

$$\therefore b^2 - 4ac = 0 \tag{4.14}$$

$$\therefore 4(l\alpha + m\beta + n\gamma)^2 = 4\left(l^2 + m^2 + n^2\right)\left(\alpha^2 + \beta^2 + \gamma^2 - a^2\right)$$

$$\therefore (l\alpha + m\beta + n\gamma)^2 = \left(l^2 + m^2 + n^2\right)\left(\alpha^2 + \beta^2 + \gamma^2 - a^2\right)$$

is the condition for the line (4.11) to touch a given sphere.

Eliminating $l, m,$ and n by replacing by $(x - \alpha); (y - \beta)$ and $(z - \gamma)$ respectively

$$\therefore [\alpha(x-\alpha) + \beta(y-\beta) + \gamma(z-\gamma)]^2$$
$$= [(x-\alpha)^2 + (y-\beta)^2 + (z-\gamma)^2][\alpha^2 + \beta^2 + \gamma^2 - a^2] \tag{4.15}$$

$$\therefore [\alpha x + \beta y + \gamma z - (\alpha^2 + \beta^2 + \gamma^2)]^2 = [(x^2 + y^2 + z^2) - 2(x\alpha + \beta y + \gamma z)$$
$$- (\alpha^2 + \beta^2 + \gamma^2)][\alpha^2 + \beta^2 + \gamma^2 - a^2].$$

Let S: $x^2 + y^2 + z^2 - a^2$; $S_1 : \alpha^2 + \beta^2 + \gamma^2 - a^2$ and

$$T : \alpha x + \beta y + \gamma z - a^2$$

$$\therefore [(T + a^2) - (S_1 + a^2)]^2 = [(S + a^2) - 2(T + a^2) + (S + a^2)]S_1$$

$$\therefore (T - S_1)^2 = (S - 2T + S_1)\,S_1$$

$$\therefore T^2 - 2T\,S_1 + S_1^2 = S\,S_1 - 2T\,S_1 + S_1^2$$

$$\therefore T^2 = SS_1^2$$

which is the required equation of the enveloping cone.

6) Find the enveloping cone of the sphere $x^2 + y^2 + z^2 - 2x + 4z = 1$ with its vertex at $(1, 1, 1)$.

Sol. Let $S : x^2 + y^2 + z^2 - 2x + 4z - 1 = 0$.

$$S_1 : (S)_{(1,1,1)} = 1 + 1 + 1 - 2 + 4 - 1 = 4.$$

$$T : (1)x + (1)y + (1)z - (x + 1) + 2(z + 1) - 1 = y + 3z.$$

$\therefore$ Equation of cone is

$$SS_1 = T^2$$

$$\therefore (x^2 + y^2 + z^2 - 2x - 4z - 1)(4) = (y + 3z)^2$$

$$\therefore 4x^2 + 4y^2 + 4z^2 - 8x + 16z - 4 = y^2 + 6yz + 9z^2$$

$$\therefore 4x^2 + 3y^2 - 5z^2 - 6yz - 8x + 16z - 4 = 0.$$

7) Show that the plane $z = 0$ cuts the enveloping cone of the sphere $x^2 + y^2 + z^2 = 11$ which has its vertex at $(2, 4, 1)$ in a rectangular hyperbola.

Sol. The equation of the sphere is $x^2 + y^2 + z^2 - 11 = 0$ and the vertex is $(2, 4, 1)$.

Let $S : x^2 + y^2 + z^2 - 11$ at $(x_1, y_1, z_1) = (2, 4, 1)$

$$S_1 : x_1^2 + y_1^2 + z_1^2 - 11 = 4 + 16 + 1 - 11 = 10$$

$$T : xx_1 + yy_1 + zz_1 - 11 = 2x + 4y + z - 11.$$

The equation of the enveloping cone is $SS_1 = T^2$

$$\therefore (x^2 + y^2 + z^2 - 11)(10) = (2x + 4y + z - 11)^2$$

4.3 Enveloping Cone to a Surface

which meets the curve $z = 0$.

$$\therefore 10(x^2 + y^2 - 11) - (2x + 4y - 11)^2 = 0$$
$$\therefore 10x^2 + 10y^2 - 110 - (4x^2 + 16y^2 + 121 + 16xy - 88y - 44x) = 0$$
$$\therefore 6x^2 - 6y^2 + 16xy - 88y - 44x - 231 = 0.$$

which represents a rectangular hyperbola in the XY plane because of the coefficient of x^2 + coefficient of $y^2 = 0$.

8) The sections of the enveloping cone of the surface $\frac{x^2}{a^2} + \frac{y^2}{b^2} + \frac{z^2}{c^2} = 1$ whose vertex is $P(\alpha, \beta, \gamma)$ by the plane $z = 0$ is a rectangular hyperbola, Find the locus of the vertex P.

Sol. The equation of the given surface is

$$S: \frac{x^2}{a^2} + \frac{y^2}{b^2} + \frac{z^2}{c^2} - 1 = 0, \quad S_1: \frac{\alpha^2}{a^2} + \frac{\beta^2}{b^2} + \frac{\gamma^2}{c^2} - 1 = 0$$

and $T = \frac{\alpha x}{a^2} + \frac{\beta y}{b^2} + \frac{\gamma z}{c^2} = 1$.

$\therefore$ Equation of the enveloping cone with vertex $P(\alpha, \beta, \gamma)$ is $SS_1 = T^2$

$$\therefore \left(\frac{x^2}{a^2} + \frac{y^2}{b^2} + \frac{z^2}{c^2} - 1 \right) \left(\frac{\alpha^2}{a^2} + \frac{\beta^2}{b^2} + \frac{\gamma^2}{c^2} - 1 \right) = \left(\frac{\alpha x}{a^2} + \frac{\beta y}{b^2} + \frac{\gamma z}{c^2} - 1 \right)^2$$

it meets plane $z = 0$.

$$\therefore \left(\frac{x^2}{a^2} + \frac{y^2}{b^2} - 1 \right) \left(\frac{\alpha^2}{a^2} + \frac{\beta^2}{b^2} + \frac{\gamma^2}{c^2} - 1 \right) = \left(\frac{\alpha x}{a^2} + \frac{\beta y}{b^2} - 1 \right)^2. \quad (4.16)$$

Equation (4.16) represents a rectangular hyperbola if the coefficient of x^2 + coefficient of $y^2 = 0$

$$\Rightarrow \frac{1}{a^2} \left(\frac{\beta^2}{b^2} + \frac{\gamma^2}{c^2} - 1 \right) + \frac{1}{b^2} \left(\frac{\alpha^2}{a^2} + \frac{\gamma^2}{c^2} - 1 \right) = 0$$

$$\Rightarrow \frac{\alpha^2 + \beta^2}{a^2 b^2} + \frac{\gamma^2}{c^2} \left(\frac{a^2 + b^2}{a^2 b^2} \right) = \frac{a^2 + b^2}{a^2 b^2}$$

$$\therefore \frac{\alpha^2 + \beta^2}{a^2 + b^2} + \frac{\gamma^2}{c^2} = 1$$

Hence the locus of (α, β, γ) is $\frac{x^2 + y^2}{a^2 + b^2} + \frac{z^2}{c^2} = 1$.

4.4 Equation of the Cone whose Vertex is the Origin is Homogeneous

Theorem:

The equation of the cone whose vertex is the origin is homogeneous and conversely.

or Every second-degree equation in x, y, z represents a cone with vertex at the origin if and only if it is second degree homogeneous in $x, y,$ and z.

Proof: The general second-degree equation is

$$ax^2 + by^2 + cz^2 + 2hxy + 2fyz + 2gzx + 2ux + 2vy + 2wz + d = 0. \quad (4.17)$$

Now, to prove that it represents a cone with its vertex origin if and only if

$$u = v = w = d = 0.$$

Suppose Equation (4.17) represents a cone with a vertex at the origin. Let $P(x_1, y_1, z_1)$ be a point on the cone represented with vertex at $O(0,0,0)$.
∴ Direction ratios of OP passing through the point P will be

$$(x_1 - 0;\ y_1 - 0;\ z_1 - 0) = (x_1, y_1, z_1).$$

∴ Equation of generator of the cone will be

$$\frac{x-0}{x_1} = \frac{y-0}{y_1} = \frac{z-0}{z_1} = t \quad (4.18)$$
$$\therefore x = x_1 t;\ y = y_1 t;\ z = z_1 t.$$

Let $Q(x_1 t, y_1 t, z_1 t)$ lie on the cone.
∴ It must satisfy the Equation (4.17).

$$\therefore a(x_1 t)^2 + b(y_1 t)^2 + c(z_1 t)^2 + 2h(x_1 t)(y_1 t) + 2f(y_1 t)(z_1 t)$$
$$+ 2g(z_1 t)(x_1 t) + 2u(x_1 t) + 2v(y_1 t) + 2w(z_1 t) + d = 0$$
$$\therefore t^2 \left[ax_1^2 + by_1^2 + cz_1^2 + 2hx_1 y_1 + 2fy_1 z_1 + 2gz_1 x_1 \right]$$
$$+ 2t \left[ux_1 + vy_1 + wz_1 \right] + d = 0 \quad (4.19)$$

which is true for all values of t; $t \neq 0$.

$$\therefore ax_1^2 + by_1^2 + cz_1^2 + 2hx_1 y_1 + 2fy_1 z_1 + 2gz_1 x_1 = 0;\ ux_1 + vy_1 + wz_1 = 0;\ d = 0.$$

4.4 Equation of the Cone whose Vertex is the Origin is Homogeneous

From equation $ux_1 + vy_1 + wz_1 = 0$; if u, v, w are not all zero then the coordinates x_1, y_1, z_1 of any point on the cone satisfy an equation of the first degree. i.e., $ux + vy + wz = 0$ which is the equation of the plane which contradicts our hypothesis that $P(x_1, y_1, z_1)$ lies on a cone.

∴ u, v, w all are not zero is false.
∴ $u = 0 = v = w$ and by equation $d = 0$.
∴ General second-degree equation reduces to

$$ax^2 + by^2 + cz^2 + 2fyz + 2gzx + 2hxy = 0$$

which is a homogeneous second-degree equation.

Converse:

Let
$$ax^2 + by^2 + cz^2 + 2hxy + 2gzx + 2fyz = 0, \quad (4.20)$$

be a given second homogeneous equation in x, y, z then to prove that it represents a cone with vertex origin.

Let $P(x_1, y_1, z_1)$ be any point that satisfies the Equation (4.20)

∴ $ax_1^2 + by_1^2 + cz_1^2 + 2h(x_1)(y_1) + 2g(z_1)(x_1) + 2f(y_1)(z_1) = 0$

∴ $t^2 \left[ax_1^2 + by_1^2 + cz_1^2 + 2hx_1y_1 + 2gx_1z_1 + 2fy_1z_1 \right] = 0; t \neq 0$

∴ $a(x_1t)^2 + b(y_1t)^2 + c(z_1t)^2 + 2h(x_1t)(y_1t) + 2g(x_1t)(z_1t)$
$\qquad + 2f(y_1t)(z_1t) = 0.$

∴ Point $Q(x_1t, y_1t, z_1t)$ satisfies the Equation (4.20).

i.e., If $P(x_1, y_1, z_1)$ satisfies the Equation (4.20) then the point (x_1t, y_1t, z_1t) also satisfies the Equation (4.20).

∴ If P lies on the surface represented by (4.20) then every point Q on the line $\overrightarrow{OP}$ lies on it.

∴ The surface is generated by line through the origin and by definition is a cone with its vertex at O.

Corollary:

If l, m, n be the direction ratios of any generator of a cone with vertex origin then (l, m, n) satisfies the equation of the cone.

Proof: Let the equation of cone with vertex origin be

$$ax^2 + by^2 + cz^2 + 2hxy + 2gzx + 2fyz = 0. \quad (4.21)$$

Equation of generator through vertex origin and having direction ratios l, m, n is

$$\frac{x-0}{l} = \frac{y-0}{m} = \frac{z-0}{n} = t$$

$$\therefore x = lt;\ y = mt;\ z = nt$$

which satisfies the Equation (4.21).

$$\therefore a(lt)^2 + b(mt)^2 + c(nt)^2 + 2h(lt)(mt) + 2g(nt)(lt)$$

$$+ 2f(mt)(nt) = 0$$

$$\therefore t^2[al^2 + bm^2 + cn^2 + 2hlm + 2gln + 2fmn] = 0$$

$$\therefore al^2 + bm^2 + cn^2 + 2hlm + 2gln + 2fmn = 0\ (\because t \neq 0)$$

which proves that (l, m, n) satisfy the Equation (4.21).

9) Find the equation of the cone whose vertex is at the origin and which passes through the curve given by the equations

$$ax^2 + by^2 + cz^2 = 1;\ lx + my + nz = p.$$

Sol. The equation of the cone with vertex origin is a second-degree homogeneous equation.

$$\therefore lx + my + nz = p$$

$$\therefore \frac{lx+my+nz}{p} = 1$$

$\therefore$ The required equation of cone is

$$ax^2 + by^2 + cz^2 = (1)^2$$

$$\therefore ax^2 + by^2 + cz^2 = \left(\frac{lx+my+nz}{p}\right)^2$$

$$\therefore p^2\left(ax^2 + by^2 + cz^2\right) = (lx + my + nz)^2.$$

10) Find the equation of the cone whose vertex is at the origin and the direction cosines of whose generators satisfy the relation $3l^2 - 4m^2 + 5n^2 = 0$.

Sol. The vertex of the given cone is the origin

$\therefore$ The equation of any generator through the origin with direction cosines l, m, n are

$$\frac{x}{l} = \frac{y}{m} = \frac{z}{n} \qquad (4.22)$$

4.4 Equation of the Cone whose Vertex is the Origin is Homogeneous

Also satisfies the relation

$$3l^2 - 4m^2 + 5n^2 = 0. \tag{4.23}$$

Direction cosines of a generator of a cone with vertex at origin satisfy the equation of the cone

∴ Equation (4.23) becomes, $3x^2 - 4y^2 + 5z^2 = 0$
which is the required equation of the cone.

11) Find the equation of the cone with vertex at the origin and which pass through the curves $z = 2$ and $x^2 + y^2 = 4$.

Sol. The equation of a cone with vertex origin is a second-degree homogeneous curve.

$$\therefore \frac{z}{2} = 1$$

∴ the required equation of cone is

$$x^2 + y^2 = 4(1)$$
$$\therefore x^2 + y^2 = 4\left(\frac{z}{2}\right)^2$$
$$\therefore x^2 + y^2 - z^2 = 0.$$

12) Find the equation of the cone with vertex origin and passes through the curve $x^2 + y^2 + z^2 - x - 1 = 0$ and $x^2 + y^2 + z^2 + y - 2 = 0$.

Sol. The equations of the guiding curve are

$$x^2 + y^2 + z^2 - x - 1 = 0, \tag{4.24}$$

and

$$x^2 + y^2 + z^2 + y - 2 = 0. \tag{4.25}$$

Let us assume t, for making both the equations homogeneous in x, y and z
∴ Equation (4.24) becomes

$$x^2 + y^2 + z^2 - tx - t^2 = 0, \tag{4.26}$$

and Equation (4.25) becomes

$$x^2 + y^2 + z^2 + ty - 2t^2 = 0. \tag{4.27}$$

Eliminating t from Equations (4.26) and (4.27), we get

$$-tx - ty + t^2 = 0$$
$$\therefore -t(x + y - t) = 0$$
$$\therefore x + y = 0 (\because t \neq 0)$$

Substituting the value of t in Equation (4.26), we get

$$x^2 + y^2 + z^2 - x(x+y) - (x+y)^2 = 0$$
$$\therefore -x^2 + z^2 - 3xy = 0$$
$$\therefore x^2 - z^2 + 3xy = 0$$

which is the required equation of the cone.

13) Find the equation of the cone with vertex at the origin and which pass through the curves $x^2 + y^2 + z^2 + 2x - 3y + 4z = 5$.

Sol. Given curve is the intersection of the two spheres

$$S_1 : x^2 + y^2 + z^2 + x - 2y + 3z - 4 = 0$$
$$S_2 : x^2 + y^2 + z^2 + 2x - 3y + 4z - 5 = 0. \qquad (4.28)$$

$\therefore$ Equation of the plane of the circle of their intersection is

$$S_1 - S_2 = 0$$
$$\therefore -x + y - z + 1 = 0$$
$$\therefore x - y + z = 1.$$

$\therefore$ The equation of a cone with vertex origin is second-degree homogeneous.
$\therefore$ Equation (4.28) becomes

$$x^2 + y^2 + z^2 + (x - 2y + 3z)(1) - 4(1)^2 = 0$$
$$\therefore (x^2 + y^2 + z^2) + (x - 2y + 3z)(x - y + z) - 4(x - y + z)^2 = 0$$
$$\therefore x^2 + y^2 + z^2 + x^2 - xy + xz - 2xy + 2y^2 - 2yz + 3xz - 3yz + 3z^2$$
$$- 4x^2 - 4y^2 - 4z^2 + 8xy + 8yz - 8xz = 0$$
$$\therefore -2x^2 - y^2 + 5xy + 3yz - 4xz = 0$$
$$\therefore 2x^2 + y^2 - 5xy - 3yz + 4xz = 0$$

is the required equation of the cone.

14) Show that the general equation to a cone that passes through the three axes is $fyz + gzx + hxy = 0$ where f, g, h are parameters.

4.4 Equation of the Cone whose Vertex is the Origin is Homogeneous

Sol. We know that generator lines of any cone intersect only in its vertex.
Here coordinate axes are generators and they intersect in origin.
∴ A required cone is a cone with vertex origin.
∴ Equation of cone must be of the form

$$ax^2 + by^2 + cz^2 + 2hxy + 2gzx + 2fyz = 0. \qquad (4.29)$$

Let the x axis be the generator.
∴ Its direction cosine will satisfy the equation of cone i.e., $(1,0,0)$ satisfies equation (4.29).

$$\therefore a = 0.$$

Similarly, the y axis and z axis are generators.
∴ $(0,1,0)$ and $(0,0,1)$ respectively are direction cosines of the y axis and z axis which satisfies the Equation (4.29).
∴ b = 0 and c = 0
∴ Equation (4.29) becomes,

$$2fyz + 2gzx + 2hxy = 0$$

$$\therefore fyz + gzx + hxy = 0$$

is the required equation of cone having three coordinate axes as generators where $f, g,$ and h are parameters.

15) Find the equation to the cone which passes through the three coordinate axes as well as the two lines $\frac{x}{1} = \frac{y}{-2} = \frac{z}{3}$ and $\frac{x}{3} = \frac{y}{-1} = \frac{z}{1}$.

Sol. Equation of the cone passes through the three coordinate axes is

$$fyz + gzx + hxy = 0 \qquad (4.30)$$

The equation of lines are $\frac{x}{1} = \frac{y}{-2} = \frac{z}{3}$ and $\frac{x}{3} = \frac{y}{-1} = \frac{z}{1}$ which are generators of cone.
∴ Direction ratios are $(1, -2, 3)$ and $(3, -1, 1)$
∴ Equation (4.30) becomes,

$$f(-2)(3) + g(3)(1) + h(1)(-2) = 0$$
$$\therefore -6f + 3g - 2h = 0, \qquad (4.31)$$

and $f(-1)(1) + g(1)(3) + h(3)(-1) = 0$

$$\therefore -f + 3g - 3h = 0 \qquad (4.32)$$

Solving (4.31) and (4.32), we get $f = \frac{h}{5}$.
Substituting in (4.32), we get $g = \frac{16h}{15}$
∴ Equation (4.30) becomes; $\left(\frac{h}{5}\right)yz + \left(\frac{16}{15}h\right)zx + hxy = 0$.
∴ $3yz + 16zx + 15xy = 0$ is the required equation of the cone.

16) Find the equation of the quadric cone which passes through the three coordinate axes and the three mutually perpendicular lines

$$\frac{x}{2} = y = -z;\ x = \frac{y}{3} = \frac{z}{5}\ \text{and}\ \frac{x}{8} = \frac{-y}{11} = \frac{z}{5}.$$

Sol. The equation of any cone passing through the coordinate axes is

$$fyz + gzx + hxy = 0 \qquad (4.33)$$

It passes through the lines

$$\frac{x}{2} = y = -z \Rightarrow \frac{x}{2} = \frac{y}{1} = \frac{z}{-1}$$

and

$$x = \frac{y}{3} = \frac{z}{5} \Rightarrow \frac{x}{1} = \frac{y}{3} = \frac{z}{5}.$$

∴ Direction ratios are $(2, 1, -1)$ and $(1, 3, 5)$ with satisfies Equation (4.33)

$$\therefore f(1)(-1) + g(-1)(2) + h(2)(1) = 0$$
$$\therefore -f - 2g + 2h = 0, \qquad (4.34)$$

and $f(3)(5) + g(5)(1) + h(1)(3) = 0$

$$\therefore 15f + 5g + 3h = 0 \qquad (4.35)$$

Solving (4.34) and (4.35), we get

$$\frac{f}{\begin{vmatrix} -2 & 2 \\ 5 & 3 \end{vmatrix}} = \frac{g}{\begin{vmatrix} 2 & -1 \\ 3 & 15 \end{vmatrix}} = \frac{h}{\begin{vmatrix} -1 & -2 \\ 15 & 5 \end{vmatrix}}$$

$$\therefore \frac{f}{-6-10} = \frac{g}{30+3} = \frac{h}{-5+30}$$

$$\therefore \frac{f}{-16} = \frac{g}{33} = \frac{h}{25}$$

∴ $f = \frac{-16h}{25}$ and $g = \frac{33h}{25}$
∴ Equation (4.33) becomes;

$$\frac{-16h}{25}yz + \frac{33h}{25}zx + hxy = 0$$
$$\therefore 16yz - 33zx - 25xy = 0 \qquad (4.36)$$

which will be the required cone if it satisfied the third given line $\frac{x}{8} = \frac{-y}{11} = \frac{z}{5}$
∴ The direction ratios are $(8, -11, 5)$.
L.H.S. of Equation (4.36) $= 16yz - 33zx - 25xy$
$$= 16(-11)(5) - 33(5)(8) - 25(8)(-11)$$
$$= -880 - 1320 + 2200$$
$$= 0 = \text{R.H.S.}$$

∴ Line $\frac{x}{8} = \frac{-y}{11} = \frac{z}{5}$ also satisfies the equation of the cone.
∴ $16yz - 33zx - 25xy = 0$ is the required equation of the cone.

4.5 Intersection of a Line with a Cone

To find the points of intersection of the line $\frac{x-\alpha}{l} = \frac{y-\beta}{m} = \frac{z-\gamma}{n}$ and cone with vertex origin.

Let the equation of the cone with vertex origin is

$$ax^2 + by^2 + cz^2 + 2fz + 2gzx + 2hxy = 0. \qquad (4.37)$$

And the equation of the line is

$$\frac{x-\alpha}{l} = \frac{y-\beta}{m} = \frac{z-\gamma}{n} = t. \qquad (4.38)$$

∴ Coordinates of the line (4.38) are $x = \alpha + lt; y = \beta + mt; z = \gamma + nt$
Substituting the values of x, y, and z in Equation (4.37), we get

$$a(\alpha + lt)^2 + b(\beta + mt)^2 + c(\gamma + nt)^2 + 2f(\beta + mt)(\gamma + nt)$$
$$+ 2g(\gamma + nt)(\alpha + lt) + 2h(\alpha + lt)(\beta + mt) = 0,$$

∴ $t^2[al^2 + bm^2 + cn^2 + 2fmn + 2gln + 2hlm] + 2t[aal + bm\beta + c\gamma n$
$+ \gamma fm + f\beta n + g\gamma l + g\alpha n + h\alpha m + hl\beta] + [a\alpha^2 + b\beta^2 + c\gamma^2 + 2f\beta\gamma$
$+ 2g\gamma\alpha + 2h\alpha\beta] = 0$
∴ $t^2[al^2 + bm^2 + cn^2 + 2fmn + 2gln + 2hlm] + 2t[(a\alpha + h\beta +$
$g\gamma)l + (h\alpha + b\beta + f\gamma)m + (g\alpha + f\beta + c\gamma)n] + [a\alpha^2 + b\beta^2$
$+ c\gamma^2 + 2f\beta\gamma + 2g\gamma\alpha + 2h\alpha\beta] = 0$
∴ $t^2\left[al^2 + bm^2 + cn^2 + 2fmn + 2gnl + 2hlm\right]$
$+ 2t[l(a\alpha + h\beta + g\gamma) + m(h\alpha + b\beta + f\gamma)$
$+ n(g\alpha + f\beta + c\gamma)] + f(\alpha, \beta, \gamma) = 0$

150 Cone

which is a quadratic equation in t.
∴ It has roots t_1 and t_2.
∴ Points of intersection of line and cone are
$P(\alpha + lt_1, \beta + mt_1, \gamma + nt_1)$ and $Q(\alpha + lt_2; \beta + mt_2; \gamma + nt_2)$.

4.6 Equation of a Tangent Plane at (α, β, γ) to the Cone with Vertex Origin

Let the equation line through a point (α, β, γ) of the cone be

$$\frac{x-\alpha}{l} = \frac{y-\beta}{m} = \frac{z-\gamma}{n}. \tag{4.39}$$

The equation of cone with vertex origin is

$$f(x, y, z) = ax^2 + by^2 + cz^2 + 2fyz + 2gzx + 2hxy = 0 \tag{4.40}$$

$$\therefore f(\alpha, \beta, \gamma) = a\alpha^2 + b\beta^2 + c\lambda^2 + 2f\beta\gamma + 2g\gamma\alpha + 2h\alpha\beta = 0. \tag{4.41}$$

Let l, m, n be the direction ratio of a line through the point (α, β, γ)
∴ Coordinates of the line (4.39) are

$$x = \alpha + lt; y = \beta + mt; z = \gamma + nt. \tag{4.42}$$

Substitution value of $x, y,$ and z from Equation (4.40), we get

$$\therefore t^2[al^2 + bm^2 + cn^2 + 2fmn + 2gnl + 2hlm] + 2t[l(a\alpha + h\beta + g\gamma)$$
$$+ m(h\alpha + b\beta + f\gamma) + n(g\alpha + f\beta + c\gamma)]$$
$$+ [a\alpha^2 + b\beta^2 + c\gamma^2 + 2f\beta\gamma + 2g\gamma\alpha + 2h\alpha\beta] = 0$$

$$\begin{aligned} t^2[f(l, m, n)] + 2t[l(a\alpha + h\beta + g\gamma) + m(h\alpha + b\beta + f\gamma) \\ + n(g\alpha + f\beta + c\gamma)] + f(\alpha, \beta, \gamma) = 0. \end{aligned} \tag{4.43}$$

∴ Equation (4.39) is a tangent line to cone (4.40) if and only if quadratic Equation (4.41) has two equal roots.

$$\text{i.e., } \Delta = 0$$

$$\text{i.e., } b^2 - 4ac = 0$$

$$\therefore b^2 = 4ac$$

4.6 Equation of a Tangent Plane at (α, β, γ) to the Cone with Vertex Origin

i.e., $[2(l(a\alpha + h\beta + g\gamma) + m(h\alpha + b\beta + f\gamma) + n(g\alpha + f\beta + c\gamma))]^2$
$- 4f(l, m, n) f(\alpha, \beta, \gamma) = 0$

$\therefore l(a\alpha + h\beta + g\gamma) + m(h\alpha + b\beta + f\gamma) + n(g\alpha + f\beta + c\gamma) = 0$ (4.44)

$(\because f(\alpha, \beta, \gamma) = 0;$ By (4.41))

Eliminating l, m, n from Equations (4.39) and (4.44); we get

i.e., Replacing l, m, n by $(x - \alpha); (y - \beta)$ and $(z - \gamma)$ respectively in (4.44); we get

$\therefore (x - \alpha)(a\alpha + h\beta + g\gamma) + (y - \beta)(h\alpha + b\beta + f\gamma)$
$+ (z - \gamma)(g\alpha + f\beta + c\gamma) = 0$
$\therefore x(a\alpha + h\beta + g\gamma) + y(h\alpha + b\beta + f\gamma) + z(g\alpha + f\beta + c\gamma)$
$= \alpha(a\alpha + h\beta + g\gamma) + \beta(h\alpha + b\beta + f\gamma) + \gamma(g\alpha + f\beta + c\gamma)$
$= a\alpha^2 + b\beta^2 + c\gamma^2 + 2f\beta\gamma + 2g\gamma\alpha + 2h\alpha\beta = f(\alpha, \beta, \gamma)$

which is the required equation of the tangent plane.

Remark :

To remember the equation of the tangent plane

	α	β	γ
x	a	h	g
y	h	b	f
z	g	f	c

$= x(a\alpha + h\beta + g\gamma) + y(h\alpha + b\beta + f\gamma) + z(g\alpha + f\gamma + c\gamma).$

Corollary:

The tangent plane at any point $(k\alpha, k\beta, k\gamma)$ on the generator through the point (α, β, γ) is the same as the tangent plane at (α, β, γ).

Proof: Equation of the tangent plane at $(k\alpha, k\beta, k\gamma)$ to the cone with vertex origin is given by

$x[a(k\alpha) + h(k\beta) + g(k\gamma)] + y[h(k\alpha) + b(k\beta) + f(k\gamma)]$
$+ z[g(k\alpha) + f(k\beta) + c(k\gamma)] = 0$
$\therefore k[x(a\alpha + h\beta + g\gamma) + y(h\alpha + b\beta + f\gamma) + z(g\alpha + f\beta + c\gamma)] = 0$
$\therefore x(a\alpha + h\beta + g\gamma) + y(h\alpha + b\beta + f\gamma) + z(g\alpha + f\beta + c\gamma) = 0$

which is the required equation of tangent plane at point (α, β, γ).

$\therefore$ The tangent plane at any point on the cone touches the cone at all points of the generator through that point and we say that the plane touches the cone along with the generator.

4.7 Conditions for Tangency

To find the condition that the plane $lx + my + nz = 0$ should touch the cone
$$ax^2 + by^2 + cz^2 + 2hxy + 2fyz + 2gzx = 0.$$

Proof: Let the equation of the plane be
$$lx + my + nz = 0. \tag{4.45}$$

The equation of the cone is
$$ax^2 + by^2 + cz^2 + 2hxy + 2fyz + 2gzx = 0. \tag{4.46}$$

Let (α, β, γ) be the point of contact to the tangent plane
$$x(a\alpha + h\beta + g\gamma) + y(h\alpha + b\beta + f\gamma) + z(g\alpha + f\beta + c\gamma) = 0. \tag{4.47}$$

$\therefore$ By Equations (4.45) and (4.47) planes are identical if and only if
$$\frac{a\alpha + h\beta + g\gamma}{l} = \frac{h\alpha + b\beta + f\gamma}{m} = \frac{g\alpha + f\beta + c\gamma}{n} = k$$

$$\left.\begin{array}{l} \therefore a\alpha + h\beta + g\gamma - kl = 0 \\ h\alpha + b\beta + f\gamma - mk = 0 \\ g\alpha + f\beta + c\gamma - nk = 0 \end{array}\right\} \tag{4.48}$$

$\therefore$ Point (α, β, γ) lies on the plane (4.45); we get
$$l\alpha + m\beta + n\gamma = 0 \tag{4.49}$$

Eliminating α, β, γ, k from (4.48) and (4.49), we get
$$\begin{vmatrix} a & h & g & l \\ h & b & f & m \\ g & f & c & n \\ l & m & n & o \end{vmatrix} = 0$$

which is the required condition for tangency.
The determinant on expansion gives
$$Al^2 + Bm^2 + Cn^2 + 2Fmn + 2Gln + 2Hlm = 0,$$

where A, B, C, F, G and H are the cofactors of a, b, c, f, g, h respectively in determinant

$$\begin{vmatrix} a & h & g \\ h & b & f \\ g & f & c \end{vmatrix}$$

where $A = bc - f^2; B = ac - g^2; C = ab - h^2$

$$H = fg - hc; F = hg - af; G = hf - bg.$$

17) Find the plane which touches the cone $x^2 + 2y^2 - 3z^2 + 2yz - 5zx + 3xy = 0$ along with the generator whose direction ratios are $1, 1, 1$.

Sol. The required plane touches the cone $x^2 + 2y^2 - 3z^2 + 2yz - 5zx + 3yx = 0$.

Comparing with the general equation

$$ax^2 + by^2 + cz^2 + 2fyz + 2gzx + 2hxy = 0$$
$$\therefore a = 1; b = 2; c = -3; f = 1; g = \frac{-5}{2}; h = \frac{3}{2}$$

Direction ratios of the generator are $1, 1, 1$
$\therefore$ Equation of generators are $\frac{x}{1} = \frac{y}{1} = \frac{z}{1} = r$.
$\therefore$ Any point on this generator is (r, r, r).
The tangent plane at the point (r, r, r) to the given cone is

$$x(a\alpha + h\beta + g\gamma) + y(h\alpha + b\beta + f\gamma) + z(g\alpha + f\beta + c\gamma) = 0$$
$$\therefore x\left(1r + \frac{3}{2}r - \frac{5}{2}r\right) + y\left(\frac{3}{2}r + 2r + r\right) + z\left(\frac{-5}{2}r + r - 3r\right) = 0$$
$$\therefore x\left(1 + \frac{3}{2} - \frac{5}{2}\right) + y\left(\frac{3}{2} + 2 + 1\right) + z\left(\frac{-5}{2} + 1 - 3\right) = 0$$
$$\therefore \frac{9}{2}y - \frac{9}{2}z = 0$$
$$\therefore y = z$$

is the required tangent plane.

18) Prove that the perpendiculars drawn from the origin to the tangent planes to the tangent planes to the cone $ax^2 + by^2 + cz^2 = 0$ lie on the cone

$$\frac{x^2}{a} + \frac{y^2}{b} + \frac{z^2}{c} = 0.$$

Sol. The perpendicular drawn from the origin to the tangent planes to the given cone is called the **reciprocal cone**.

Given the equation of the cone is $ax^2 + by^2 + cz^2 = 0$.
∴ By conditions for tangency

$$Ax^2 + By^2 + Cz^2 + 2Fyz + 2Gzx + 2Hxy = 0,$$

where,

$$A = bc + f^2 = bc;\ B = ca - g^2 = ac;\ C = ab - h^2 = ab;$$

$$F = 0 = G = H$$

$$\therefore bcx^2 + cay^2 + abz^2 = 0$$

$$\therefore \frac{x^2}{a} + \frac{y^2}{b} + \frac{z^2}{c} = 0$$

is the required equation of the cone.

19) Prove that tangent planes to the cone $x^2 - y^2 + 2z^2 - 3yz + 4zx - 5xy = 0$ are perpendicular to the generators of the cone $17x^2 + 8y^2 + 29z^2 + 28yz - 46zx - 16xy = 0$.

Sol. Given the equation of cone is

$$x^2 - y^2 + 2z^2 - 3yz + 4zx - 5xy = 0$$

Comparing with the general equation of cone

$$ax^2 + by^2 + cz^2 + 2fyz + 2gx + 2hxy = 0$$

$$\therefore a = 1, b = -1, c = 2, 2f = -3, 2g = 4, 2h = -5, f = \frac{-3}{2},$$

$$g = 2, h = \frac{-5}{2}.$$

By condition of tangency

$$Ax^2 + By^2 + Cz^2 + 2Fyz + 2Gzx + 2Hxy = 0. \qquad (4.50)$$

$$\therefore A = bc - f^2 = (-1)(2) - \left(\frac{-3}{2}\right)^2 = \frac{-17}{4}$$

$$B = ca - g^2 = (2)(1) - (2)^2 = -2$$

$$C = ab - h^2 = 1(-1) - \left(\frac{-5}{2}\right)^2 = \frac{-29}{4}$$

$$F = gh - af = 2\left(\frac{-5}{2}\right) - 1\left(\frac{-3}{2}\right) = \frac{-7}{2}$$

$$G = hf - bg = \frac{-5}{2}\left(\frac{-3}{2}\right) - (-1)(2) = \frac{23}{4}$$

$$H = fg - ch = \left(\frac{-3}{2}\right)(2) - 2\left(\frac{-5}{2}\right) = 2.$$

∴ Equation (4.50) becomes,

$$\frac{-17x^2}{4} - 2y^2 - \frac{29z^2}{4} + 2\left(\frac{-7}{2}\right)yz + 2\left(\frac{23}{4}\right)zx + 2(2)xy = 0$$
$$\therefore 17x^2 + 8y^2 + 29z^2 + 28yz - 46zx - 16xy = 0$$

which is the required equation of the cone.

20) Find the equation of the cone which passes through the common generators of the cones $-2x^2 + 4y^2 + z^2 = 0$ and $10xy - 2yz + 5zx = 0$ and the line with direction cosines proportional to $1, 2, 3$.

Sol. Any cone passing through the common generators of the cone

$$-2x^2 + 4y^2 + z^2 = 0 \text{ and } 10xy - 2yz + 5zx = 0$$

is

$$(-2x^2 + 4y^2 + z^2) + \lambda(10xy - 2yz + 5zx) = 0. \qquad (4.51)$$

Equation (4.51) passes through a line whose direction ratios are $1, 2, 3$.
  ∴ Direction ratios satisfy the Equation (4.51)

$$\therefore (-2 + 16 + 9) + \lambda(20 - 12 + 15) = 0$$
$$\therefore \lambda = -1.$$

∴ Substituting in Equation (4.51), we get

$$2x^2 - 4y^2 - z^2 + 10xy - 2yz + 5zx = 0$$

which is the required equation of the cone.

21) Show that the equation of the cone through the intersection of the cones $x^2 - 2y^2 + 3z^2 - 4yz + 5zx - 6xy = 0$ and $2x^2 - 3y^2 + 4z^2 + 5yz + 6zx + 10xy = 0$ and the line with direction cosines proportional to $1, 1, 1$ is $y^2 - 2z^2 + 3yz - 4zx + 2xy = 0$.

Sol. Let the equation of the cone through the intersection of the cones
$$x^2 - 2y^2 + 3z^2 - 4yz + 5zx - 6xy = 0,$$
and
$$2x^2 - 3y^2 + 4z^2 - 5yz + 6zx - 10xy = 0$$
is
$$\left(x^2 - 2y^2 + 3z^2 - 4yz + 5zx - 6xy\right)$$
$$+\gamma \left(2x^2 - 3y^2 + 4z^2 - 5yz + 6zx - 10xy\right) = 0. \quad (4.52)$$

Equation (4.52) passes through the line with direction ratios 1, 1, 1
∴ Direction cosines satisfy the Equation (4.52)

$$\therefore [1 - 2 + 3 - 4 + 5 - 6] + \gamma [2 - 3 + 4 - 5 + 6 - 10] = 0$$
$$\therefore -3 - 6\gamma = 0$$
$$\therefore \gamma = \frac{-1}{2}.$$

∴ Equation (4.52) becomes,
$$\left(x^2 - 2y^2 + 3z^2 - 4yz + 5zx - 6xy\right)$$
$$-\frac{1}{2}\left(2x^2 - 3y^2 + 4z^2 + 5yz + 6zx - 10xy\right) = 0$$
$$\therefore 2x^2 + 4y^2 + 6z^2 - 8yz + 10zx - 12xy - 2x^2 + 3y^2 - 4z^2 + 5yz - 6zx + 10xy = 0$$
$$\therefore -y^2 + 2z^2 - 3yz + 4zx - 2xy = 0$$
$$\therefore y^2 - 2z^2 + 3yz - 4zx + 2xy = 0$$

is the required equation of the cone.

4.8 Right Circular Cone

A right circular cone is a surface generated by a line that passes through a fixed point and makes a constant angle with a fixed line through the fixed point.

The fixed point is called the vertex.

The fixed-line through the vertex is called the axis and the fixed angle is called the semi-vertical angle of the cone and the straight line through the vertex is the generator.

Theorem: Every section of a right circular cone by a line perpendicular to its axis is a circle.

Proof:

Let a plane perpendicular to the axis on a right circular cone with a semi-vertical angle α meets it at N.

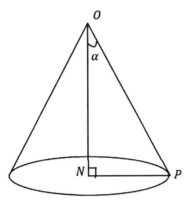

Let P be a point of the section; since ON is perpendicular to the plane which contains the line NP.

$$\therefore ON \perp NP$$

$$\therefore \frac{PN}{ON} = \tan \alpha$$

$$\therefore NP = ON \tan \alpha .$$

NP is constant for every position of point P of the section.

$\therefore$ The section is circular with N as its center.

$\therefore$ Every section of a right circular cone perpendicular to the axis is a circle.

Equation of a right circular cone:

To find the equation of cone with axis as a line $\frac{x-\alpha}{l} = \frac{y-\beta}{m} = \frac{z-\gamma}{n}$ and vertex is the point (α, β, γ) and θ be the semi-vertical angle.

Proof:

Let $O(\alpha, \beta, \gamma)$ be the vertex and OA be the axis of the cone. The required equation is to be obtained by using the condition that the line joining any point $P(x, y, z)$ on the curve to the vertex $O(\alpha, \beta, \gamma)$ makes an angle θ with the axis OA.

$\therefore$ Direction cosines of $\overleftrightarrow{OP}$ is $x - \alpha; y - \beta; z - \gamma$

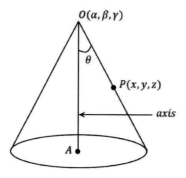

∴ The equation of axis OA is $\frac{x-\alpha}{l} = \frac{y-\beta}{m} = \frac{z-\gamma}{n}$,
where l, m, n are proportional to direction cosines θ is the angle between $\overrightarrow{OA}$ and $\overrightarrow{OP}$.

$$\therefore \cos\theta = \frac{l(x-\alpha) + m(y-\beta) + n(z-\gamma)}{\left(\sqrt{l^2+m^2+n^2}\right)\left(\sqrt{(x-\alpha)^2 + (y-\beta)^2 + (z-\gamma)^2}\right)}.$$

∴ Required equation of right circular cone is

$$[l(x-\alpha) + m(y-\beta) + n(z-\gamma)]^2$$
$$= (l^2+m^2+n^2)\left[(x-\alpha)^2 + (y-\beta)^2 + (z-\gamma)^2\right]\cos^2\theta.$$

Corollary 1: If the vertex be the origin; then the equation of the right circular cone is

$$(lx + my + nz)^2 = (l^2+m^2+n^2)(x^2+y^2+z^2)\cos^2\theta.$$

Corollary 2: If the vertex be the origin and the axis of the cone be the z-axis then $(\alpha, \beta, \gamma) = (0, 0, 0)$ and $(l, m, n) = (0, 0, 1)$.
∴ Equation of right circular cone becomes,

$$z^2 = (x^2+y^2+z^2)\cos^2\theta$$
$$\therefore x^2+y^2+z^2 = \frac{z^2}{\cos^2\theta}$$
$$\therefore x^2+y^2 = z^2\sec^2\theta - z^2$$
$$\therefore x^2+y^2 = z^2(\sec^2\theta - 1)$$
$$\therefore x^2+y^2 = z^2\tan^2\theta.$$

4.8 Right Circular Cone

22) Show that the equation of the right circular cone which passes through the point $(1, 1, 2)$ and has its vertex at origin and axis the line $\frac{x}{2} = \frac{-y}{4} = \frac{z}{3}$ is $4x^2 + 40y^2 + 19z^2 - 48xy - 72yz + 36xz = 0$.

Sol. The axis of the cone is $\frac{x}{2} = \frac{y}{-4} = \frac{z}{3}$.

Let $P(1, 1, 2)$ be any point on the cone.
The direction ratio of generator OP are $1, 1, 2$.
Let α be the semi-vertical angle of the cone

$$\therefore \cos\alpha = \frac{2(1)+(-4)(1)+3(2)}{\sqrt{2^2+(-4)^2+(3)^2}\sqrt{1^2+1^2+2^2}}$$

$$\therefore \cos\alpha = \frac{4}{\sqrt{174}}.$$

Let $Q(x, y, z)$ be any point on the surface of the cone.
In $\triangle OQM$, $\cos\alpha = \frac{OM}{OQ}$

$$\therefore OQ^2 \cos^2\alpha = OM^2. \tag{4.53}$$

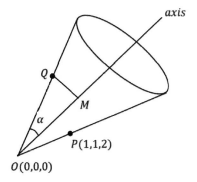

$\therefore OM =$ Projection of OP on-axis whose direction ratios are $2, -4, 3$.
$\therefore$ Direction cosines are $\frac{2}{\sqrt{29}}; \frac{-4}{\sqrt{29}}; \frac{3}{\sqrt{29}}$.

$$\therefore OM = \frac{2x}{\sqrt{29}} - \frac{4y}{\sqrt{29}} + \frac{3z}{\sqrt{29}}$$
$$OQ^2 = x^2 + y^2 + z^2.$$

$\therefore$ Equation (4.53) becomes,

$$(x^2+y^2+z^2)\left(\frac{4}{\sqrt{174}}\right)^2 = \left(\frac{2x-4y+3z}{\sqrt{29}}\right)^2$$
$$\therefore (16)(29)(x^2+y^2+z^2) = 174(2x-4y+3z)^2$$
$$\therefore 8(x^2+y^2+z^2) = 3(4x^2+16y^2+9z^2-16xy+12xz-2yz)$$
$$\therefore 4x^2+40y^2+19z^2-48xy+36xz-72yz = 0.$$

23) Show that the equation of the right circular cone with vertex (2, 3, 1) axis parallel to the line $-x = \frac{y}{2} = z$ and one of its generators having direction cosines proportional to (1, -1, 1) is

$$x^2 - 8y^2 + z^2 - 12xy - 12yz + 6zx - 46x + 36y + 22z - 19 = 0.$$

Sol. The axis of the cone is the line through vertex $A(2, 3, 1)$ and parallel to NL.

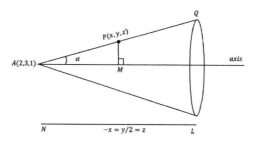

∴ Its equations are

$$\frac{x-2}{-1} = \frac{y-3}{2} = \frac{z-1}{1}$$

∴ It has direction ratios $(-1, 2, 1)$, or $(1, -2, -1)$
Direction ratios of one generator are $(1, -1, 1)$.
Let α be the semi-vertical angle of the cone

$$\therefore \cos\alpha = \frac{(1)(1)+(-2)(-1)+(-1)(1)}{\sqrt{1+4+1}\sqrt{1+1+1}}$$
$$\therefore \cos\alpha = \frac{2}{\sqrt{18}} = \frac{2}{3\sqrt{2}} = \frac{\sqrt{2}}{3}.$$

Let PM be perpendicular to the axis of the cone.
In $\triangle APM$; $\cos\alpha = \frac{AM}{AP}$

$$\therefore AP^2 \cos^2\alpha = AM^2. \tag{4.54}$$

AM = Projection of AP on the axis of the cone

$$= (x-2)\left(\frac{1}{\sqrt{6}}\right) + (y-3)\left(\frac{-2}{\sqrt{6}}\right) + (z-1)\left(\frac{-1}{\sqrt{6}}\right)$$
$$\therefore AM = \frac{x-2y-z+5}{\sqrt{6}}.$$

AP = Distance between points A and P

$$\therefore AP^2 = (x-2)^2 + (y-3)^2 + (z-1)^2$$

$\therefore$ Equation (4.54) becomes,

$$\left[(x-2)^2 + (y-3)^2 + (z-1)^2\right]\frac{2}{9} = \left[\frac{x-2y-z+5}{\sqrt{6}}\right]^2$$

$$\therefore 4\left[x^2 - 4x + 4 + y^2 - 6y + 9 + z^2 - 2z + 1\right]$$
$$= 3\left[x^2 + 4y^2 + z^2 + 25 - 4xy - 2xz + 10x + 4yz\right] - 20y - 10z$$
$$\therefore -x^2 + 8y^2 - z^2 - 12xy - 6xz + 12yz + 46x - 36y - 22z + 19 = 0$$
$$\therefore x^2 - 8y^2 + z^2 + 12xy + 6xz - 12yz - 46x + 36y + 22z - 19 = 0$$

is the required equation of the cone.

24) Lines are drawn through the origin with direction cosines proportional to (1, 2, 2); (2, 3, 6); (3, 4, 12). Show that the axis of the right circular cone through then has direction cosines $\left(\frac{-1}{\sqrt{3}}, \frac{1}{\sqrt{3}}, \frac{1}{\sqrt{3}}\right)$ and that the semi-vertical angle of the cone is $\cos^{-1}\left(\frac{1}{\sqrt{3}}\right)$ obtain also the equation of the cone.

Sol. Let α be the semi-vertical angle of the cone and l, m, n be the direction ratios of the axis of the cone.

The lines with direction ratios (1, 2, 2); (2, 3, 6), and (3, 4, 12) are the generators of the cone

$$\therefore \cos\alpha = \frac{l + 2m + 2n}{\sqrt{l^2 + m^2 + n^2}\sqrt{1^2 + 2^2 + 2^2}}$$

$$\cos\alpha = \frac{2l + 3m + 6n}{\sqrt{l^2 + m^2 + n^2}\sqrt{2^2 + 3^2 + 6^2}}$$

$$\cos\alpha = \frac{3l + 4m + 12n}{\sqrt{l^2 + m^2 + n^2}\sqrt{3^2 + 4^2 + 12^2}}$$

$$\therefore \frac{l + 2m + 2n}{3\sqrt{l^2 + m^2 + n^2}} = \frac{2l + 3m + 6n}{7\sqrt{l^2 + m^2 + n^2}}$$

$$\therefore 7(l + 2m + 2n) = 3(2l + 3m + 6n)$$

$$\therefore l + 5m - 4n = 0. \qquad (4.55)$$

Similarly, $\dfrac{l+2m+2n}{3\sqrt{l^2+m^2+n^2}} = \dfrac{3l+4m+12n}{13\sqrt{l^2+m^2+n^2}}$.

$\therefore 13(l+2m+2n) = 3(3l+4m+12n)$
$\therefore 2l+7m-5n = 0$ (4.56)

$\therefore$ By (4.55) and (4.56), we get

$$\dfrac{l}{\begin{vmatrix} 5 & -4 \\ 7 & -5 \end{vmatrix}} = \dfrac{m}{\begin{vmatrix} -41 \\ -52 \end{vmatrix}} = \dfrac{n}{\begin{vmatrix} 15 \\ 27 \end{vmatrix}}$$

$\therefore \dfrac{l}{-25+28} = \dfrac{m}{-8+5} = \dfrac{n}{7-10}$

$\therefore \dfrac{l}{3} = \dfrac{m}{-3} = \dfrac{n}{-3}$.

$\therefore$ Direction ratios of the axis of the cone are (-1, 1, 1).
$\therefore$ Direction Cosines of the cone are $\left(\dfrac{-1}{\sqrt{3}}, \dfrac{1}{\sqrt{3}}, \dfrac{1}{\sqrt{3}}\right)$.

$\therefore \cos\alpha = \dfrac{1(-1)+2(1)+2(1)}{\sqrt{3}\cdot(3)} = \dfrac{1}{\sqrt{3}}$

$\therefore \alpha = \cos^{-1}\left(\dfrac{1}{\sqrt{3}}\right)$.

Let $P(x, y, z)$ be any point on the cone.
Let $\overline{PM}$ be perpendicular to the axis of the cone.

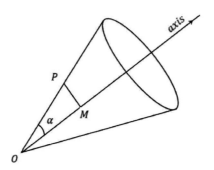

$\therefore OM = \dfrac{-x}{\sqrt{3}} + \dfrac{y}{\sqrt{3}} + \dfrac{z}{\sqrt{3}}$

$OP = \sqrt{x^2+y^2+z^2}$

$\therefore OM^2 = OP^2\cos^2\alpha$

4.8 Right Circular Cone

$$\therefore \left[\frac{-x+y+z}{\sqrt{3}}\right]^2 = \left(\frac{1}{\sqrt{3}}\right)^2 (x^2+y^2+z^2)$$

$$\therefore x^2+y^2+z^2 - 2xy - 2xz + 2yz = x^2+y^2+z^2$$

$$\therefore xy - yz + zx = 0$$

is the required equation of the cone.

25) If α is the semi-vertical angle of the right circular cone that passes through the lines. $OY; OZ; x=y=z$; Show that $\cos\alpha = (9-4\sqrt{3})^{-1/2}$.

Sol. Cone passes through origin and generators of the cone are OY, OZ;

$$x=y=z.$$

$\therefore$ Direction ratios are $(0, 1, 0); (0, 0, 1)$, and $(1, 1, 1)$.

Let l, m, n be the direction ratios of the axis of the cone. Let α be the semi-vertical angle of the right circular cone.

$$\therefore \cos\alpha = \frac{l(0)+m(1)+n(0)}{\sqrt{l^2+m^2+n^2}\sqrt{0+1+0}} = \frac{m}{\sqrt{l^2+m^2+n^2}}, \quad (4.57)$$

$$\cos\alpha = \frac{l(0)+m(0)+n(1)}{\sqrt{l^2+m^2+n^2}\sqrt{0+0+1}} = \frac{n}{\sqrt{l^2+m^2+n^2}}, \quad (4.58)$$

$$\cos\alpha = \frac{l(1)+m(1)+n(1)}{\sqrt{l^2+m^2+n^2}\sqrt{1+1+1}} = \frac{l+m+n}{\sqrt{3(l^2+m^2+n^2)}}. \quad (4.59)$$

$\therefore$ By (4.57) and (4.58) $\Rightarrow m = n$

$$\Rightarrow m - n = 0. \quad (4.60)$$

Solving (4.59) and (4.60)

$$\frac{l}{(-1)(1-\sqrt{3})-1} = \frac{m}{1(0)-1(-1)} = \frac{n}{(1)(1)-0}$$

$$\frac{m}{1} = \frac{n}{1} = \frac{l}{\sqrt{3}-2}$$

$$\therefore \cos\alpha = \frac{1}{\sqrt{(\sqrt{3}-2)^2+1^2+1^2}}$$

$$= \frac{1}{\sqrt{3+4-4\sqrt{3}+2}}$$

$$\therefore \cos\alpha = \frac{1}{\sqrt{9-4\sqrt{3}}}$$

$$\therefore \cos\alpha = (9-4\sqrt{3})^{-1/2}.$$

Cone

Exercise:

1) Find the equations to the cones with vertex at the origin and passes through the curve

   (a) $x^2 + y^2 + z^2 - x - 1 = 0, x^2 + y^2 + z^2 + y - 2 = 0$
   (b) $ax^2 + by^2 + cz^2 - 1 = 0, \alpha x^2 + \beta y^2 - 2z = 0$
   (c) $\frac{x^2}{a^2} + \frac{y^2}{b^2} = 1, z = c$

   Answer: (a) $x^2 - 3xy - z^2 = 0$,
   (b) $4z^2 \left(ax^2 + by^2 + cz^2\right) = \alpha \left(x^2 + \beta y^2\right)^2$,
   (c) $\frac{x^2}{a^2} + \frac{y^2}{b^2} - \frac{z^2}{c^2} = 0$

2) Find the equation of the cone with a vertex is $(1, 2, 3)$ and base is $y^2 = 4ax, z = 0$.

   Answer: $(2z - 3y)^2 = 4a (z - 3) (z - 3x)$

3) Find the equation of the cone with vertex $(5, 4, 3)$, and $3x^2 + 2y^2 = 6, y + z = 0$ as a base.

   Answer: $147x^2 + 87y^2 + 101z^2 - 210xy + 90yz - 210zx - 294 = 0$

4) Show that the line $\frac{x}{l} = \frac{y}{m} = \frac{z}{n}$, where $l^2 + 3m^2 - 3n^2 = 0$ is a generator of the cone $x^2 + 2y^2 - 3z^2 = 0$.

5) Prove that the perpendiculars are drawn from the origin to tangent planes to the cone $2x^2 + 3y^2 + 4z^2 + 2yz + 4zx + 6xy = 0$ lie on the cone $11x^2 + 4y^2 - 3z^2 + 8yz - 6zx - 20xy = 0$.

6) Show that the locus of the lines of intersection of tangent planes to the cone $ax^2 + by^2 + cz^2 = 0$ which touch along perpendicular generators is cone $a^2 (b + c) x^2 + b^2 (c + a) y^2 + c^2 (a + b) z^2 = 0$.

7) Find the equation to the right circular cone whose vertex is $(2, -3, 5)$, axis PQ which makes equal angles with the axes and semi-vertical angle is $30°$.

   Answer: $5x^2 + 5y^2 + 5z^2 - 8xy - 8zx - 4x + 86y - 58z + 278 = 0$

8) Find the equation to the right circular cone whose vertex is $P(2, -3, 5)$, axis PQ which makes equal angles with the axes, and which passes through the point $A(1, -2, 3)$.

Answer: $x^2 + y^2 + z^2 + 6(xy + yz + zx) - 16x - 36y - 4z - 28 = 0$

9) Find the equation of the right circular cone whose vertex is at the origin, whose axis is the line $\frac{x}{1} = \frac{y}{2} = \frac{z}{3}$, and which has a vertical angle of $60°$.

Answer: $19x^2 + 13y^2 + 3z^2 - 8xy - 24yz - 12zx = 0$

10) Prove that $x^2 - y^2 + z^2 - 4x + 2y + 6z + 12 = 0$ represents a right circular cone whose vertex is the point $(2, 1, -3)$ whose axis is parallel to OY and whose semi-vertical angle is $45°$.

11) Find the equation of the right circular cone, whose vertex is the origin and semi-vertical angle is $45°$ and axis is $x = y = z$.

Answer: $x^2 + y^2 + z^2 = 4(xy + yz + zx)$.

5

Cylinder

5.1 Definition

A cylinder is a surface generated by straight lines parallel to a fixed-line and satisfying one more condition intersecting to a curve or touching a given surface.

The intersecting curve is known as the guiding curve. Straight lines are known as generators and the fixed line is known as the axis.

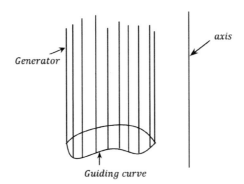

Remark:

The minimum requirement of the cylinder is

1) Direction ratio of the axis $L\ (l, m, n)$

2) Guiding curve $\begin{cases} S = 0 \\ u = 0 \end{cases}.$

5.2 Equation of the Cylinder whose Generators Intersect the Given Conic and are Parallel to a Given Line

To find the equation of the cylinder whose generators intersect the conic $ax^2 + 2hxy + by^2 + 2gx + 2fy + c = 0$, $z = 0$ and parallel to the line $\frac{x}{l} = \frac{y}{m} = \frac{z}{n}$.

Sol. Let $P(\alpha, \beta, \gamma)$ be any point on the cylinder parallel to the line $\frac{x}{l} = \frac{y}{m} = \frac{z}{n}$ having direction ratio (l, m, n).

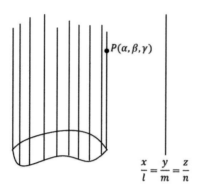

$\therefore$ Equation of generator through $P(\alpha, \beta, \gamma)$ are

$$\frac{x-\alpha}{l} = \frac{y-\beta}{m} = \frac{z-\gamma}{n}. \qquad (5.1)$$

Since line (5.1) intersect the conic

$$\begin{cases} ax^2 + 2hxy + by^2 + 2gx + 2fy + c = 0 \\ z = 0 \end{cases}$$

$\therefore$ Line (5.1) will be satisfied by $z = 0$

$$\therefore \frac{x-\alpha}{l} = \frac{y-\beta}{m} = \frac{0-\gamma}{n}$$

$$\therefore x = \alpha - \frac{l\gamma}{n};\ y = \beta - \frac{m\gamma}{n},\ \text{and}\ z = 0$$

are general coordinates if line (5.1) which satisfy the surface of the conic S.

$$ax^2 + 2hxy + by^2 + 2gx + 2fy + c = 0$$

$$\therefore a\left(\alpha - \frac{l\gamma}{n}\right)^2 + 2h\left(\alpha - \frac{l\gamma}{n}\right)\left(\beta - \frac{m\gamma}{n}\right) + b\left(\beta - \frac{m\gamma}{n}\right)^2 + 2g\left(\alpha - \frac{l\gamma}{n}\right)$$

5.2 Equation of the Cylinder whose Generators Intersect the Given Conic

$$+2f\left(\beta - \frac{m\gamma}{n}\right) + c = 0$$

$$\therefore a(\alpha n - l\gamma)^2 + 2h(\alpha n - l\gamma)(\beta n - m\gamma) + b(\beta n - m\gamma)^2$$
$$+ 2gn(\alpha n - l\gamma) + 2fn(\beta n - m\gamma) + cn^2 = 0$$

which is the condition that the point (α, β, γ) should lie on the surface of the cylinder.

$\therefore$ For the equation of the cylinder replace (α, β, γ) by the point (x, y, z).

$$\therefore a(nx - lz)^2 + 2h(nx - lz)(ny - mz) + b(ny - mz)^2$$
$$+ 2gn(nx - lz) + 2fn(ny - mz) + cn^2 = 0$$

which is the required equation of cylinder.

Remark:

If the generator is parallel to the z-axis, then $l = 0$; $m = 0$ and $n = 1$ in the equation of the cylinder become $ax^2 + 2hxy + by^2 + 2gx + 2fy + c = 0$.

1) Find the equation of the cylinder whose generators are parallel to the line $x = \frac{-y}{2} = \frac{z}{3}$ and whose guiding curve is the ellipse $x^2 + 2y^2 = 1$; $z = 3$.

Sol. Generators are parallel to the line $\frac{x}{1} = \frac{-y}{2} = \frac{z}{3}$ and the guiding curve is the ellipse $x^2 + 2y^2 = 1$; $z = 3$.

Let $P(x_1, y_1, z_1)$ be any point on the cylinder, then the equation of any generator through the point P is

$$\frac{x - x_1}{1} = \frac{y - y_1}{-2} = \frac{z - z_1}{3}$$

$$\therefore \frac{x - x_1}{1} = \frac{y - y_1}{-2} = \frac{3 - z_1}{3}$$

$$\therefore \frac{x - x_1}{1} = \frac{y - y_1}{-2} = 1 - \frac{z_1}{3}$$

$$\therefore x = x_1 + 1 - \frac{z_1}{3}; y = -2 + \frac{2z_1}{3} + y_1; z = 3.$$

This generator intersects the given conic if

$$\left(x_1 + 1 - \frac{z_1}{3}\right)^2 + 2\left(y_1 + \frac{2z_1}{3} - 2\right)^2 = 1.$$

Generalizing x_1, y_1, z_1; we get $\left(x - \frac{z}{3} + 1\right)^2 + 2\left(y + \frac{2z}{3} - 2\right)^2 = 1$

$$\therefore x^2 + 2y^2 + z^2 - \frac{2}{3}xz + \frac{8}{3}yz + 2x - 8y - 6z + 8 = 0$$

$$\therefore 3(x^2 + 2y^2 + z^2) + 2(4yz - xz) + 6(x - 4y - 3z) + 24 = 0.$$

170 Cylinder

2) Find the equation of the cylinder whose generators intersect the curve $ax^2 + by^2 = 2z;\ lx + 3y + nz = p$ and are parallel to the z-axis.

Sol. The given curves are

$$ax^2 + by^2 = 2z \tag{5.2}$$

$$lx + my + nz = p. \tag{5.3}$$

Eliminating z from (5.2) and (5.3), we get

$$ax^2 + by^2 = 2z$$
$$= \frac{2}{n}[p - lx - my]$$
$$\therefore n\left(ax^2 + by^2\right) = 2p - 2lx - 2my$$
$$\therefore n(ax^2 + by^2) + 2lx + 2my - 2p = 0.$$

5.3 Enveloping Cylinder

The locus of the tangents to a sphere that is parallel to a given line is a cylinder known as the enveloping cylinder.

Straight lines are known as generators and the fixed line is known as the axis of an enveloping cylinder.

Equation of the enveloping cylinder:

To find the equation of the cylinder whose generator touch the sphere $x^2 + y^2 + z^2 = a^2$ and are parallel to the line $\frac{x}{l} = \frac{y}{m} = \frac{z}{n}$.

Let $P(\alpha, \beta, \gamma)$ be any point on a generator parallel to a fixed-line with direction ratio (l, m, n).

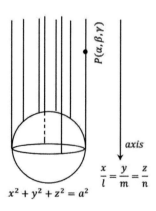

∴ Equation of generator through point $P(\alpha, \beta, \gamma)$ is

$$\frac{x-\alpha}{l} = \frac{y-\beta}{m} = \frac{z-\gamma}{n} = t, \tag{5.4}$$

$$\therefore x = \alpha + lt;\ y = \beta + mt;\ z = \gamma + nt \tag{5.5}$$

are generator coordinates of the line (5.4).

Since generator (5.4) touches the given sphere

$$x^2 + y^2 + z^2 = a^2 \tag{5.6}$$

∴ Substituting general coordinates in Equation (5.6), we get

$$\therefore (\alpha + lt)^2 + (mt + \beta)^2 + (nt + \gamma)^2 = a^2$$
$$\therefore t^2 (l^2 + m^2 + n^2) + 2t(l\alpha + m\beta + \gamma n) + (\alpha^2 + \beta^2 + \gamma^2 - a^2) = 0.$$

∴ Straight line (5.4) touches the sphere if and only if the quadratic equation must have equal roots.

i.e., $\Delta = 0$
$$\therefore b^2 - 4ac = 0$$
$$\therefore b^2 = 4ac$$

$$\therefore [2(l\alpha + m\beta + n\gamma)] = 4(l + m + n)(\alpha + \beta + \gamma - a)$$
$$\therefore (l\alpha + m\beta + n\gamma)^2 = (l^2 + m^2 + n^2)(\alpha^2 + \beta^2 + \gamma^2 - a^2).$$

This is the condition for a line to touch the sphere at (α, β, γ).

Replacing α, β, γ by x, y, z respectively, we get

$$\therefore (lx + my + nz)^2 = (l^2 + m^2 + n^2)(x^2 + y^2 + z^2 - a^2)$$

which is the required equation of enveloping cylinder.

3) Find the enveloping cylinder of the sphere $x^2 + y^2 + z^2 - 2x + 4y = 1$ having its generators parallel to the line $x = y = z$.

Sol. Let $P(\alpha, \beta, \gamma)$ be any point on the enveloping cylinder.

∴ Equation of the generator through $P(\alpha, \beta, \gamma)$ is $\frac{x-\alpha}{1} = \frac{y-\beta}{1} = \frac{z-\gamma}{1} = t$.

$$\therefore x = t + \alpha;\ y = t + \beta;\ z = t + \gamma.$$

This point satisfies the equation of the sphere

$$x^2 + y^2 + z^2 - 2x + 4y = 1$$
$$\therefore (t+\alpha)^2 + (t+\beta)^2 + (t+\gamma)^2 - 2(t+\alpha) + 4(t+\beta) = 1$$
$$\therefore 3t^2 + 2t(\alpha + \beta + \gamma + 1) + (\alpha^2 + \beta^2 + \gamma^2 - 2\alpha + 4\beta - 1) = 0.$$

172 Cylinder

Since this generator is a tangent to a given sphere.
∴ The quadratic equation must have equal roots

$$\therefore \Delta = 0$$
$$\therefore b^2 - 4ac = 0$$
$$\therefore b^2 = 4ac$$

$\therefore [2(\alpha + \beta + \gamma + 1)]^2 = 4(3)(\alpha^2 + \beta^2 + \gamma^2 - 2\alpha + 4\beta - 1)$
$\therefore \alpha^2 + \beta^2 + \gamma^2 + 1 + 2\alpha\beta + 2\alpha\gamma + 2\alpha + 2\beta\gamma + 2\beta + 2\gamma$
$\qquad = 3\alpha^2 + 3\beta^2 + 3\gamma^2 - 6\alpha + 12\beta - 3$
$\therefore 2\alpha^2 + 2\beta^2 + 2\gamma^2 - 2\alpha\beta - 2\alpha\gamma - 2\beta\gamma - 8\alpha + 10\beta - 2\gamma - 4 = 0$
$\therefore \alpha^2 + \beta^2 + \gamma^2 - \alpha\beta - \alpha\gamma - \beta\gamma - 4\alpha + 5\beta - \gamma - 2 = 0.$

Replace α, β and γ by $x, y,$ and z respectively, we get

$$x^2 + y^2 + z^2 - xy - xz - yz - 4x + 5y - z - 2 = 0$$

is the required enveloping cylinder of the sphere.

4) Find the equation of the enveloping cylinder of the coincide $\frac{x^2}{a^2} + \frac{y^2}{b^2} + \frac{z^2}{c^2} = 1$ whose generators are parallel to the line $\frac{x}{l} = \frac{y}{m} = \frac{z}{n}$.

Sol. Let $P(x_1, y_1, z_1)$ be any point on the enveloping cylinder then the equations of the generator through $P(x_1, y_1, z_1)$ are

$$\frac{x - x_1}{l} = \frac{y - y_1}{m} = \frac{z - z_1}{n} = r.$$

Any point on this generator is $(x_1 + lr, y_1 + mr, z_1 + nr)$.
If the point lies on the given conicoid, we get

$$\left(\frac{x_1 + lr}{a^2}\right)^2 = \left(\frac{y_1 + mr}{b^2}\right)^2 = \left(\frac{z_1 + nr}{c^2}\right)^2 = 1$$

$$\therefore r^2 \left(\frac{l^2}{a^2} + \frac{m^2}{b^2} + \frac{n^2}{c^2}\right) + 2r\left(\frac{lx_4}{a^2} + \frac{my_1}{b^2} + \frac{nz_1}{c^2}\right) + \left(\frac{x_1^2}{a^2} + \frac{y_1^2}{b^2} + \frac{z_1^2}{c^2}\right) = 0.$$
(5.7)

Since this generator is a tangent to the given corticoid so the two values of r obtained from (5.7) must be equal.

$$\therefore \left(\frac{lx_1}{a^2} + \frac{my_1}{b^2} + \frac{nz_1}{c^2}\right) = \left(\frac{l^2}{a^2} + \frac{m^2}{b^2} + \frac{n^2}{c^2}\right)\left(\frac{x_1^2}{a^2} + \frac{y_1^2}{b^2} + \frac{z_1^2}{c^2} - 1\right).$$

∴ The locus of $P(x_1, y_1, z_1)$

$$\left(\frac{lx}{a^2} + \frac{my}{b^2} + \frac{nz}{c^2}\right)^2 = \left(\frac{l^2}{a^2} + \frac{m^2}{b^2} + \frac{n^2}{c^2}\right)\left(\frac{x^2}{a^2} + \frac{y^2}{b^2} + \frac{z^2}{c^2} - 1\right)$$

is the required equation of the enveloping cylinder.

5) Prove that the enveloping cylinder of the ellipsoid $\frac{x^2}{a^2} + \frac{y^2}{b^2} + \frac{z^2}{c^2} = 1$ whose generators are parallel to the line $\frac{x}{0} = \frac{y}{\pm\sqrt{a^2-b^2}} = \frac{z}{c}$ meet the plane $z = 0$ in circles.

Sol. Let $P(x_1, y_1, z_1)$ be any point on the enveloping cylinder then the equations of the generator through $P(x_1, y_1, z_1)$ are

$$\frac{x - x_1}{0} = \frac{y - y_1}{\pm\sqrt{a^2 - b^2}} = \frac{z - z_1}{c} = r$$

Any point on this generator is $\left(x_1, y_1 \pm r\sqrt{a^2 - b^2}, z_1 + cr\right)$.
If this point lies on the given ellipsoid, we get

$$\frac{x_1}{a^2} + \frac{\left[y_1 \pm r\sqrt{a^2 + b^2}\right]^2}{b^2} + \frac{(z_1 + cr)^2}{c^2} = 1$$

$$\therefore r^2\left[\frac{a^2 - b^2}{b^2} + \frac{c^2}{c^2}\right] + 2r\left[\frac{cz_1}{c^2} \pm \frac{y_1\sqrt{a^2 - b^2}}{b^2}\right] + \left[\frac{x_1^2}{a^2} + \frac{y_1^2}{b^2} + \frac{z_1^2}{c^2} - 1\right] = 0$$

(5.8)

Since this generator is a tangent to the given ellipsoid, the two values of r obtained from (5.8) must be equal.

$$\therefore \left[\frac{z_1}{c} \pm \frac{y_1\sqrt{a^2-b^2}}{b^2}\right]^2 = \left[\frac{a^2-b^2}{b^2} + \frac{c^2}{c^2}\right]\left[\frac{x_1^2}{a^2} + \frac{y_1^2}{b^2} + \frac{z_1^2}{c^2} - 1\right]$$

$$\therefore \frac{z_1^2}{c^2} \pm \frac{2y_1 z_1\sqrt{a^2-b^2}}{b^2 c} + \frac{y_1^2(a^2-b^2)}{b^4} = \left(\frac{a^2}{b^2}\right)\left(\frac{x_1^2}{a^2} + \frac{y_1^2}{b^2} + \frac{z_1^2}{c^2} - 1\right)$$

∴ The locus of $P(x_1, y_1, z_1)$

$$\frac{z^2}{c^2} \pm \frac{2yz\sqrt{a^2 - b^2}}{b^2 c} + \frac{y^2(a^2 - b^2)}{b^4} = \left(\frac{a^2}{b^2}\right)\left(\frac{x^2}{a^2} + \frac{y^2}{b^2} + \frac{z^2}{c^2} - 1\right).$$

This meets the plane $z = 0$ in the curve

$$\frac{y^2(a^2 - b^2)}{b^4} = \frac{a^2}{b^2}\left(\frac{x^2}{a^2} + \frac{y^2}{b^2} - 1\right); \; z = 0.$$

i.e., $\frac{x^2}{b^2} + \frac{y^2}{b^2} = \frac{a^2}{b^2}; z = 0$.

$\therefore x^2 + y^2 = a^2; z = 0$ is the circle of radius a on the plane $z = 0$.

6) Show that the equation of the tangent plane at point (x_1, y_1, z_1) of the cylinder $ax^2 + 2hxy + by^2 + 2gx + 2fy + c = 0$ is $x(ax_1 + hy_1 + g) + y(hx_1 + by_1 + f) + (gx_1 + fy_1 + c) = 0$, and that it touches the cylinder at all points of the generator through the point.

Sol. Let the equation of the given cylinder be

$$ax^2 + 2hxy + by^2 + 2gx + 2fy + c = 0. \tag{5.9}$$

Equation of any tangent line through (x_1, y_1, z_1) is

$$\frac{x - x_1}{l} = \frac{y - y_1}{m} = \frac{z - z_1}{n} = r. \tag{5.10}$$

Any point on this tangent is $(lr + x_1, mr + y_1, nr + z_1)$

If it lies on the cylinder (5.9) then

$$a(x_1 + lr)^2 + 2h(x_1 + lr)(y_1 + mr) + b(y_1 + mr)^2 + 2g(x_1 + lr)$$
$$+ 2f(y_1 + mr) + c = 0$$

$\therefore r^2(al^2 + 2hlm + bm^2) + 2r\left[l(ax_1 + hy_1 + g) + m(hx_1 + by_1 + f)\right]$
$$+ ax_1^2 + 2hx_1y_1 + by_1^2 + 2gx_1 + 2fy_1 + c = 0 \tag{5.11}$$

which is a quadratic equation in r.

Since the point (x_1, y_1, z_1) lies on the cylinder (5.9),

$$ax_1^2 + 2hx_1y_1 + by_1^2 + 2gx_1 + 2fy_1 + c = 0. \tag{5.12}$$

By (5.11) and (5.12), we get

$$r^2\left(al^2 + 2hlm + bm^2\right) + 2r\left[l(ax_1 + hy_1 + g) + m(hx_1 + by_1 + f)\right] = 0 \tag{5.13}$$

which shows that one value of r is zero.

Since the line (5.10) touches the cylinder, it meets the cylinder only at one point, so both values of r in (5.13) must be equal.

Since one root is zero, the other must also be zero.

$\therefore$ Coefficient of $r = 0$.

$$\therefore l(ax_1 + hy_1 + g) + m(hx_1 + by_1 + f) = 0. \tag{5.14}$$

Eliminating l, m, n from (5.10) and (5.14), the locus of tangent lines (5.10) is

$$(x - x_1)(ax_1 + hy_1 + g) + (y - y_1)(hx_1 + by_1 + f) = 0$$
$$\therefore x(ax_1 + hy_1 + g) + y(hx_1 + by_1 + f)$$
$$= ax_1^2 + 2hx_1y_1 + by_1^2 + gx_1 + fy_1.$$

Adding $gx_1 + fy_1$ on both sides, we get

$$x(ax_1 + hy_1 + g) + y(hx_1 + by_1 + f) + (gx_1 + fy_1 + c)$$
$$= ax_1^2 + 2hx_1y_1 + by_1^2 + 2gx_1 + 2fy_1 + c$$
$$= 0. \quad \text{(By (5.12))}$$

$\therefore$ The required equation of the tangent plane at (x_1, y_1, z_1) is

$$x(ax_1 + hy_1 + g) + y(hx_1 + by_1 + f) + (gx_1 + fy_1 + c) = 0 \quad (5.15)$$

Now, z is absent from the equation of the cylinder.
$\therefore$ Its generators are parallel to the z axis.
$\therefore$ Direction cosines are $0, 0, 1$.
$\therefore$ Equations of the generator through (x_1, y_1, z_1) are $\frac{x-x_1}{0} = \frac{y-y_1}{0} = \frac{z-z_1}{1}$.

Any point on this generator is $(x_1, y_1, r + z_1)$.
$\therefore$ Equations of the tangent plane at $(x_1, y_1, r + z_1)$ to the cylinder (5.15) is

$$x(ax_1 + hy_1 + g) + y(hx_1 + by_1 + f) + (gx_1 + fy_1 + c) = 0 \quad (5.16)$$

which is the same as the Equation (5.15) for all values if r.
$\therefore$ The tangent plane at every point of (5.16) is the same.
$\therefore$ The tangent plane touches the cylinder along with the generator through that point.

5.4 Right Circular Cylinder

The right circular cylinder is a surface generated by straight lines which intersect a fixed circle and is perpendicular to the plane of the circle.

The fixed circle is known as a guiding curve.

The normal line to the plane of the circle through its center is known as the axis of the right circular cylinder.

176 Cylinder

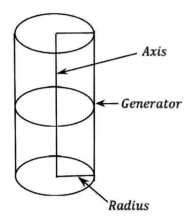

The radius of the circle is known as the radius of the cylinder.

Equation of right circular cylinder:

To find the equation of the right circular cylinder whose axis is the line $\frac{x-\alpha}{l} = \frac{y-\beta}{m} = \frac{z-\gamma}{n}$ and whose radius is r units.

Sol. Let $P(x, y, z)$ be any point on the right circular cylinder having an axis passing through $A(\alpha, \beta, \gamma)$ and with direction ratios l, m, n, and radius as r units.

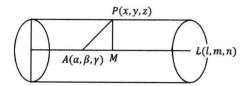

Draw PM perpendicular to axis L.
$\therefore \triangle APM$ is the right-angled triangle,

$$\therefore AP^2 = AM^2 + PM^2, \tag{5.17}$$

where $AP^2 = (x-\alpha)^2 + (y-\beta)^2 + (z-\gamma)^2$

$$PM^2 = r^2$$

5.4 Right Circular Cylinder

AM = Projection of AP on-axis L.

$$= \frac{l(x-\alpha) + m(y-\beta) + n(z-\gamma)}{\sqrt{l^2 + m^2 + n^2}}$$

$$\therefore AM^2 = \frac{[l(x-\alpha) + m(y-\beta) + n(z-\gamma)]^2}{l^2 + m^2 + n^2}.$$

∴ Equation (5.17) becomes;

$$\therefore (x-\alpha)^2 + (y-\beta)^2 + (z-\gamma)^2$$

$$= r^2 + \frac{\left[l(x-\alpha)^2 + m(y-\beta)^2 + n(z-\gamma)\right]^2}{l^2 + m^2 + n^2}$$

$$\therefore (x-\alpha)^2 + (y-\beta)^2 + (z-\gamma)^2$$

$$-\frac{\left[l(x-\alpha)^2 + m(y-\beta)^2 + n(z-\gamma)\right]^2}{l^2 + m^2 + n^2} = r^2$$

is the required equation of the right circular cylinder.

7) Find the equation of the right circular cylinder of radius 2 whose axis is the line $\frac{(x-1)}{2} = \frac{(y-2)}{1} = \left(\frac{z-3}{2}\right)$.

Sol. Let $P(x_1, y_1, z_1)$ be any point on the surface of the cylinder.
∴The length of the perpendicular from $P(x_1, y_1, z_1)$ to the given axis of the cylinder $\frac{x-1}{2} = \frac{y-2}{1} = \frac{z-3}{2}$ must be equal to the radius 2.
∴ Direction cosines of the axis are $\frac{2}{3}, \frac{1}{3}, \frac{2}{3}$.
∴[Length of the perpendicular from P to the axis]^2 = [radius]^2

$$\therefore \left|\begin{array}{cc} y_1-2 & z_1-3 \\ \frac{1}{3} & \frac{2}{3} \end{array}\right|^2 + \left|\begin{array}{cc} x_1-1 & y_1-2 \\ \frac{2}{3} & \frac{1}{3} \end{array}\right|^2 + \left|\begin{array}{cc} x_1-1 & z_1-3 \\ \frac{2}{3} & \frac{2}{3} \end{array}\right|^2 = (2)^2$$

$$\therefore \left[\frac{2}{3}(y_1-2) - \frac{1}{3}(z_1-3)\right]^2 + \left[\frac{1}{3}(x_1-1) - \frac{2}{3}(y_1-2)\right]^2$$

$$+ \left[\frac{2}{3}(x_1-1) - \frac{2}{3}(z_1-3)\right]^2 = 4$$

$$\therefore (2y_1 - z_1 - 1)^2 + (2z_1 - 2x_1 + 4)^2 + (x_1 - 2y_1 + 3)^2 = 36.$$

178 *Cylinder*

Generalizing x_1, y_1, z_1; we get

$$\therefore 4y^2 + z^2 + 1 - 4yz + 2z - 4y + 4z^2 + 4x^2 + 16 - 8xz + 16z - 16x$$

$$+ x^2 + 4y^2 + 9 - 4xy - 6y + 6x = 36$$

$$\therefore 5x^2 + 8y^2 + 5z^2 - 4xy - 4yz - 8xz + 22x - 16y - 14z - 10 = 0$$

is the required equation of the cylinder.

8) The axis of a right circular cylinder of radius 2 is a $\frac{x-1}{2} = \frac{y}{3} = \frac{z-3}{1}$ shows that its equation is $10x^2 + 5y^2 + 13z^2 - 12xy - 6yz - 4zx - 8x + 30y - 74z + 59 = 0$.

Sol. Let $P(x_1, y_1, z_1)$ be any point on the surface of the cylinder.
$\therefore$ Axis is

$$\frac{x-1}{2} = \frac{y}{3} = \frac{z-3}{1}. \qquad (5.18)$$

$\therefore$ Direction cosines are $\left(\frac{2}{\sqrt{14}}; \frac{3}{\sqrt{14}}; \frac{1}{\sqrt{14}} \right)$

$\therefore$ [Length of the perpendicular from P to the axis]^2 = [radius]^2

$$\therefore \left| \begin{array}{cc} x_1 - 1y_1 \\ \frac{2}{\sqrt{14}} \frac{3}{\sqrt{14}} \end{array} \right|^2 + \left| \begin{array}{cc} y_1 z_1 - 3 \\ \frac{3}{\sqrt{14}} \frac{1}{\sqrt{14}} \end{array} \right|^2 + \left| \begin{array}{cc} z_1 - 3x_1 - 1 \\ \frac{1}{\sqrt{14}} \frac{2}{\sqrt{14}} \end{array} \right|^2 = 4$$

$$\therefore (3x_1 - 2y_1 - 3)^2 + (y_1 - 3z_1 + 9)^2 + (2z_1 - x_1 - 5)^2 = 56$$

Generalizing x_1, y_1, z_1 and expanding, we get

$$\therefore 9x^2 + 4y^2 + 9 - 12xy - 18x + 12y + y^2 + 9z^2 + 81 - 6yz + 18y$$

$$- 54z + 4z^2 + x^2 + 25 - 4xz - 20z + 10x - 56 = 0$$

$$\therefore 10x^2 + 5y^2 + 13z^2 - 12xy - 6yz - 4zx - 8x + 30y - 74z + 59 = 0.$$

9) Find the equation of the circular cylinder whose guiding circle is $x^2 + y^2 + z^2 - 9 = 0$; $x - y + z = 3x - y + z = 3$.

Sol. The radius of a right circular cylinder is equal to the radius of the guiding curve and the axis of the cylinder is a line passing through the center of the circle and hence of the sphere and perpendicular to the plane of the circle. The radius of the sphere $=3$.

$\therefore$ The length of the perpendicular from the center $(0,0,0)$ to the given plane $x - y + z - 3 = 0$ is $\left| \frac{-3}{\sqrt{1+1+1}} \right| = \sqrt{3}$.

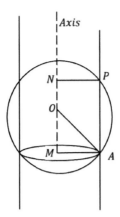

∴ The radius of the cylinder $= MA = \sqrt{OA^2 - OM^2} = \sqrt{9-3} = \sqrt{6}$.

The axis of the cylinder is the line passing through $(0,0,0)$ and normal to the plane $x - y + z = 3$.

∴ Equations are $\frac{x}{1} = \frac{y}{-1} = \frac{z}{1}$.

∴ Direction cosines are $\left(\frac{1}{\sqrt{3}}; \frac{-1}{\sqrt{3}}, \frac{1}{\sqrt{3}}\right)$.

Let $P(x, y, z)$ be any point on the cylinder.

∴ The equation of the cylinder is

$$\left(\frac{1}{\sqrt{3}}\right)^2 \left[\begin{vmatrix} y & z \\ -1 & 1 \end{vmatrix}^2 + \begin{vmatrix} z & x \\ 1 & 1 \end{vmatrix}^2 + \begin{vmatrix} x & y \\ 1 & -1 \end{vmatrix}^2\right] = \left(\sqrt{6}\right)^2$$

∴ $(y+z)^2 + (z-x)^2 + (-x-y)^2 = 18$
∴ $2x^2 + 2y^2 + 2z^2 + 2yz - 2xz + 2xy = 18$
∴ $x^2 + y^2 + z^2 + yz - xz + xy - 9 = 0$

which is the required equation of the right circular cylinder.

10) Obtain the equation of the right circular cylinder described on the circle through the three points $(1,0,0)$, $(0,1,0)$, $(0,0,1)$ as the guiding curve.

Sol. Circle ABC can be interpreted as the intersection of the sphere $OABC$ and the plane ABC; O being the origin.

∴ Equations of the circle are

$$x^2 + y^2 + z^2 - x - y - z = 0, \qquad (5.19)$$

$$x + y + z = 1. \qquad (5.20)$$

∴ The axis of the cylinder will be perpendicular to the plane (5.20).
∴ Direction ratios are $(1, 1, 1)$.
∴ Generators of the cylinder will have the direction ratios $(1, 1, 1)$.
Let (x^1, y^1, z^1) be any point on the cylinder.
∴ The equation of the generator is $\frac{x-x^1}{1} = \frac{y-y^1}{1} = \frac{z-z^1}{1} = t$
$$\therefore x = t + x^1; y = t + y^1; z = t + z^1$$
Suppose this generator meets the circle at the point $(x^1 + t; y^1 + t; z^1 + t)$.
∴ Equation (5.19) and (5.20) becomes;
$$(x^1 + t)^2 + (y^1 + t)^2 + (z^1 + t)^2 - (x^1 + t) - (y^1 + t) - (z^1 + t) = 0$$
and $t = \frac{(1 - x^1 - y^1 - z^1)}{3}$.
∴ Eliminating t we get $x^{1^2} + y^{1^2} + z^{1^2} - (x^1 y^1 + y^1 z^1 + z^1 x^1) - 1 = 0$
Generalizing x^1, y^1, and z^1, we get
∴ $x^2 + y^2 + z^2 - xy - yz - zx - 1 = 0$ which is the equation of the cylinder.

11) Find the equation of the right circular cylinder of radius 3 whose axis passes through the point $(1, 3, 4)$ and has $(1, -2, 3)$ as its direction ratios.

Sol. The direction cosines of the axis are $\frac{1}{\sqrt{14}}, \frac{-2}{\sqrt{14}}, \frac{3}{\sqrt{14}}$.
The axis of the cylinder passes through the point $(1, 3, 4)$.
∴ The equation of the axis is $\frac{x-1}{1} = \frac{y-3}{-2} = \frac{z-4}{3}$.
Let (a, b, c) be any point on the cylinder.
The length of the perpendicular from the point (a, b, c) to the axis is 3.
$$(a-1)^2 + (b-3)^2 + (c-4)^2$$
$$-\left[\frac{1}{\sqrt{14}}(a-1) - \frac{2}{\sqrt{14}}(b-3) + \frac{3}{\sqrt{14}}(c-4)\right]^2 = 9$$
$$\therefore 14\left[(a-1)^2 + (b-3)^2 + (c-4)^2\right]$$
$$-[(a-1) - 2(b-3) + 3(c-4)]^2 = 126$$
∴ $12a^2 + 10b^2 + 5c^2 - 6ac + 12bc + 4ab + 5a - 112b - 70c + 198 = 0$.
Since (a, b, c) is any point on the cylinder, the required equation of the cone is $12x^2 + 10y^2 + 5z^2 - 6xz + 12yz + 4xy + 5x - 112y - 70z + 198 = 0$.

12) Find the equation of the cylinder generated by straight lines parallel to the z axis and passing through the curve of intersection of the plane
$$4x + 3y - 2z = 5 \text{ and } 3x^2 - y^2 + 2z^2 = 1.$$

Sol. The equations of the given line are

$$4x + 3y - 2z = 5 \text{ and } 3x^2 - y^2 + 2z^2 = 1$$

The equation of the cylinder generated by straight lines parallel to the z axis, we get

$$3x^2 - y^2 + 2\left(\frac{4x + 3y - 5}{2}\right)^2 = 1$$

$$\therefore 22x^2 + 7y^2 + 24xy - 40x - 30y + 23 = 0$$

is the required equation of the cylinder.

13) Find the equation of the surface generated by a straight line which is parallel to the line $y = mx, z = nx$ and intersects the ellipse

$$\frac{x^2}{a^2} + \frac{y^2}{b^2} = 1, \ z = 0.$$

Sol. The equations of the given line are $y = mx, z = nx.$

$$\Rightarrow \frac{x}{l} = \frac{y}{m} = \frac{z}{n}.$$

Let (x_1, y_1, z_1) be any point on the surface then the equation of the generator is

$$\frac{x - x_1}{l} = \frac{y - y_1}{m} = \frac{z - z_1}{n}.$$

It intersects the ellipse $\frac{x^2}{a^2} + \frac{y^2}{b^2} = 1, \ z = 0.$

Substituting $z = 0$, we get $\frac{x - x_1}{l} = \frac{y - y_1}{m} = \frac{-z_1}{n}$

$\therefore x = x_1 - \frac{z_1}{n}$ and $y = y_1 - \frac{m_1 z_1}{n}$

$\therefore$ The point of intersection is $\left(x_1 - \frac{z_1}{n}, y_1 - \frac{m_1 z_1}{n}, 0\right).$

This will satisfy the equation of the guiding curve

$$\frac{\left(x_1 - \frac{z_1}{n}\right)^2}{a^2} + \frac{\left(y_1 - \frac{m_1 z_1}{n}\right)^2}{b^2} = 1$$

$$\therefore b^2(nx_1 - z_1)^2 + a^2(ny_1 - m_1 z_1)^2 = n^2 a^2 b^2.$$

$\therefore$ The locus of (x_1, y_1, z_1) is $b^2(nx - z)^2 + a^2(ny - mz)^2 = n^2 a^2 b^2$ is the required equation of the surface.

Cylinder

14) Prove that the equation of the right circular cylinder whose axis is $\frac{x-2}{2} = \frac{y-1}{1} = \frac{z}{3}$ and passes through the point $(0, 0, 3)$ is $10x^2 + 13y^2 + 5z^2 - 6yz - 12zx - 4xy - 36x - 18y + 30z - 135 = 0$.

Sol. The equations of the axis of the cylinder are

$$\frac{x-2}{2} = \frac{y-1}{1} = \frac{z}{3}. \qquad (5.21)$$

(Radius of cylinder)^2 = (Length of perpendicular from $(0, 0, 3)$ to the line)^2

$$= \frac{1}{(2^2 + 1^2 + 3^2)} \left[(-6)^2 + (12)^2 + 0 \right] = \frac{90}{7}$$

$\therefore$ The radius of the cylinder $= \sqrt{\frac{90}{7}}$.

Let $P(x_1, y_1, z_1)$ be any point on the cylinder then the length of the perpendicular from $P(x_1, y_1, z_1)$ to the axis (5.21) must be equal to the radius of the cylinder.

i.e., $\frac{90}{7} \left[2^2 + 1^2 + 3^2 \right] = \left[(y_1 - 1)(3) - (z_1)(1) \right]^2 + \left[2z_1 - 3(x_1 - 2) \right]^2$

$$+ \left[1(x_1 - 2) - 2(y_1 - 1) \right]^2$$

$\therefore 180 = (3y_1 - z_1 - 3)^2 + (2z_1 - 3x_1 - 6)^2 + (x_1 - 2y_1)^2$

$10x_1^2 + 13y_1^2 + 5z_1^2 - 6y_1 z_1 - 12z_1 x_1 - 4x_1 y_1 - 36x_1 - 18y_1 + 30z_1 - 135 = 0$.

$\therefore$ The locus of $P(x_1, y_1, z_1)$

$$10x^2 + 13y^2 + 5z^2 - 6yz - 12xz - 4xy - 36x - 18y + 30z - 135 = 0$$

is the required equation of the cylinder.

Exercise:

1) Find the equation of the cylinder whose generators are parallel to the line $\frac{x}{1} = \frac{y}{2} = \frac{z}{3}$ and pass through the curve $x^2 + y^2 = 16$, $z = 0$.

   Answer: $9x^2 + 9y^2 + 5z^2 - 6xz + 12yz - 144 = 0$

2) Find the equation of the cylinder with generators parallel to OZ which passes through the curve of intersection of the surfaces represented by $x^2 + y^2 + 2z^2 = 12$ and $x + y + z = 1$.

   Answer: $3x^2 + 4xy + 3y^2 - 4x - 4y - 10 = 0$

3) A straight line is always parallel to the YZ plane and intersects the curve $x^2 + y^2 = a^2, z = 0$ and $x^2 = az, y = 0$; prove that it generates the surface $x^4 y^2 = (x^2 - az)^2 (a^2 - x^2)$.

4) Find the equation of the cylinder generated by the lines parallel to the line $\frac{x}{1} = \frac{y}{-2} = \frac{z}{5}$, the guiding curve being the conic $x = 0, y^2 = 8z$.

Answer: $(y + 2x)^2 = 8(z - 5x)$

5) Find the equation of the right circular cylinder whose axis is $\frac{x}{2} = \frac{y}{3} = \frac{z}{6}$ and radius 5.

Answer: $45x^2 + 40y^2 + 13z^2 - 12xy - 36yz - 24zx + 1225 = 0$

6) Find the equation of the cylinder whose generators touch the sphere $x^2 + y^2 + z^2 - 4x + 6y + 2z - 2 = 0$ and are parallel to the line with direction vector $(2, -1, -2)$.

Answer: $5x^2 + 8y^2 + 5z^2 + 4xy - 4yz + 8zx + 36y - 18z - 99 = 0$

7) Obtain the equation of the right circular cylinder described on the circle through the three points $(1, 0, 0), (0, 1, 0), (0, 0, 1)$ as guiding circle.

Answer: $x^2 + y^2 + z^2 - xy - yz - zx - 1 = 0$

8) Find the equation of the right circular cylinder whose axis is $x - 2 = z, y = 0$ and passes through the point $(3, 0, 0)$.

Answer: $x^2 + 2y^2 + z^2 - 2xz - 4x + 2z + 3 = 0$

9) Find the equation of the enveloping cylinder of the sphere $x^2 + y^2 + z^2 - 2x + 4y = 1$, having its generators parallel to $x = y = z$.

Answer: $x^2 + y^2 + z^2 - yz - zx - xy - 2x + 7y + z - 2 = 0$

10) Find the equation of the enveloping cylinder of the conicoid $\frac{x^2}{a^2} + \frac{y^2}{b^2} + \frac{z^2}{c^2} = 1$ whose generators are parallel to the line $x = y = z$.

Answer: $\left(\frac{x}{a^2} + \frac{y}{b^2} + \frac{z}{c^2}\right)^2 = \left(\frac{1}{a^2} + \frac{1}{b^2} + \frac{1}{c^2}\right)\left(\frac{x^2}{a^2} + \frac{y^2}{b^2} + \frac{z^2}{c^2} - 1\right)$.

6

Central Conicoid

6.1 Definition

The locus of the general equation
$$ax^2 + by^2 + cz^2 + 2fyz + 2gzx + 2hxy + 2ux + 2vy + 2wz + d = 0$$
of the second degree in x, y and z is called a conicoid.

Central conicoid:

All the surfaces which have a center and three principal planes are known as central conicoid.

Cone is also a central conicoid; vertex is its center.

6.2 Intersection of a Line with the Central Conicoid

Prove that every line meets the central conicoid in two points.

Proof: Let $ax^2 + by^2 + cz^2 = 1$ be the general equation of the central conicoid which intersects the line
$$\frac{x-\alpha}{l} = \frac{y-\beta}{m} = \frac{z-\gamma}{n} = t$$

$\therefore$ Coordinates of the line are given by $x = \alpha + lt; y = \beta + mt; z = \gamma + nt$.

As the line intersects the conicoid, the coordinates satisfy the equation of the conicoid.

i.e., $a(\alpha + lt)^2 + b(\beta + mt)^2 + c(\gamma + nt)^2 = 1$

$\therefore t^2(al^2 + bm^2 + cn^2) + 2t(al\alpha + bm\beta + cn\gamma) + (a\alpha^2 + b\beta^2 + c\gamma^2 - 1) = 0,$

which is a quadratic equation in t.

$\therefore$ It must have two roots t_1 and t_2.

$\therefore$ Points are $P(\alpha + lt_1, \beta + mt_1, \gamma + nt_1)$ and
$$Q(\alpha + lt_2, \beta + mt_2, \gamma + nt_2)$$

186 Central Conicoid

are two points of intersection of the line with the central conicoid.

1) Find the points of intersection of the line

$$\frac{-1}{3}(x+5) = (y-4) = \frac{1}{7}(z-11)$$

with the conicoid $12x^2 - 17y^2 + 7z^2 = 7$.

Sol. The given line is

$$\frac{x+5}{-3} = \frac{y-4}{1} = \frac{z-11}{7} = t$$

$$\therefore x = -3t - 5; y = t + 4; z = 7t + 11$$

which satisfies the equation of the conicoid.

$$12x^2 - 17y^2 + 7z^2 - 7 = 0$$
$$\therefore 12(-3t-5)^2 - 17(t+4)^2 + 7(7t+11)^2 - 7 = 0$$
$$\therefore 12(9t^2 + 3at + 25) - 17(t^2 + 8t + 16) + 7(49t^2 + 154t + 121) - 7 = 0$$
$$\therefore 434t^2 + 1302t + 868 = 0$$
$$\therefore t^2 + 3t + 2 = 0$$
$$\therefore (t+2)(t+1) = 0$$
$$\therefore t = -2 \text{ or } t = -1.$$

$\therefore$ Required two points are
when $t = -1 \Rightarrow (-2, 3, 4)$
when $t = -2 \Rightarrow (1, 2, -3)$.

6.3 Tangent Lines and Tangent Plane at a Point

To find the equation of tangent plane at a point $P(\alpha, \beta, \gamma)$ on the central conicoid $ax^2 + by^2 + cz^2 = 1$.

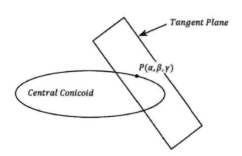

6.3 Tangent Lines and Tangent Plane at a Point

Let
$$ax^2 + by^2 + cz^2 = 1 \tag{6.1}$$
be the given central conicoid and $P(\alpha, \beta, \gamma)$ be any point on it
$$\therefore a\alpha^2 + b\beta^2 + c\gamma^2 - 1 = 0 \tag{6.2}$$
Let line passing through the point $P(\alpha, \beta, \gamma)$ be
$$\frac{x-\alpha}{l} = \frac{y-\beta}{m} = \frac{z-\gamma}{n} = t, \tag{6.3}$$
$$\therefore x = \alpha + lt; \ y = \beta + mt; \ z = \gamma + nt.$$
$\therefore$ Coordinate satisfies the equation of the conicoid
$$ax^2 + by^2 + cz^2 = 1$$
$$\therefore a(\alpha + lt)^2 + b(\beta + mt)^2 + c(\gamma + nt)^2 = 1 \tag{6.4}$$
$$\therefore t^2[al^2 + bm^2 + cn^2] + 2t[al\alpha + bm\beta + cn\gamma] + [a\alpha^2 + b\beta^2 + c\gamma^2 - 1] = 0.$$
Line (6.3) is a tangent line to a central conicoid if and only if Equation (6.4) has equal roots.

i.e., $\Delta = 0$
$$\therefore b^2 = 4ac$$
$$[2(al\alpha + bm\beta + cn\gamma)]^2 = 4[al^2 + bm^2 + cn^2][a\alpha^2 + b\beta^2 + c\gamma^2 - 1]$$
$$(\because \text{By Equation (6.2)}; a\alpha^2 + b\beta^2 + c\gamma^2 - 1 = 0)$$

which is the condition for the line to be tangent line with central conicoid.

For the equation of the tangent place replace l, m and n by $x - \alpha, y - \beta$, and $z - \gamma$ respectively.

$$\therefore a\alpha(x-\alpha) + b\beta(y-\beta) + c\gamma(z-\gamma) = 0$$
$$\therefore a\alpha x - a\alpha^2 + b\beta y - b\beta^2 + c\gamma z - c\gamma^2 = 0$$
$$\therefore a\alpha x + b\beta y + c\gamma z = a\alpha^2 + b\beta^2 + c\gamma^2$$
$$\therefore a\alpha x + b\beta y + c\gamma z = 1.$$

which is the required equation of tangent plane at point (α, β, γ) to the central conicoid.

6.4 Condition of Tangency

To find the condition that the plane $lx + my + nz = p$ is a tangent plane to a conicoid $ax^2 + by^2 + cz^2 = 1$ and also find the point of contact.

Let the given plane be

$$lx + my + nz = p, \qquad (6.5)$$

which touch the central conicoid

$$ax^2 + by^2 + cz^2 = 1. \qquad (6.6)$$

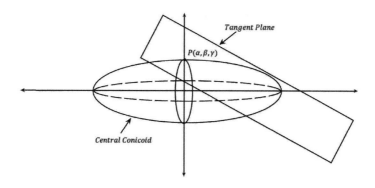

Let $P(\alpha, \beta, \gamma)$ be any point on conicoid $ax^2 + by^2 + cz^2 = 1$

$$\therefore a\alpha^2 + b\beta^2 + c\gamma^2 = 1 \qquad (6.7)$$

which is the condition that Equation (6.5) is a tangent plane.

Now, two tangent planes are given by Equations (6.7) and (6.5) which are the same if and only if their coefficients must be proportional.

$$\therefore \frac{a\alpha}{l} = \frac{b\beta}{m} = \frac{c\gamma}{n} = \frac{1}{p}$$

$$\therefore \alpha = \frac{l}{ap}; \; \beta = \frac{m}{bp}; \; \gamma = \frac{n}{cp}.$$

$\therefore$ Equation (6.7) becomes;

$$a\left(\frac{l}{ap}\right)^2 + b\left(\frac{m}{bp}\right)^2 + c\left(\frac{n}{cp}\right)^2 = 1$$

$$\therefore \frac{l^2}{a} + \frac{m^2}{b} + \frac{n^2}{c} = p^2$$

which is the required condition of tangency of the plane (6.5) to the conicoid (6.7).

The point of contact of the plane (6.5) and conicoid (6.6) is $P\left(\frac{l}{ap}; \frac{m}{bp}; \frac{n}{cp}\right)$ where $p = \pm\sqrt{\frac{l^2}{a} + \frac{m^2}{b} + \frac{n^2}{c}}$.

2) Find the equations to the tangent planes to $7x^2 - 3y^2 - z^2 + 21 = 0$ which passes through line $7x - 6y + 9 = 0$; $z = 3$.

Sol. Any plane $7x - 6y + 9 + k(z - 3) = 0$.

$$\therefore 7x - 6y + kz = 3k - 9$$

through the given line will touch given surface $7x^2 - 3y^2 - z^2 + 21 = 0$

$$\therefore \frac{-1}{3}x^2 + \frac{1}{7}y^2 + \frac{z^2}{21} = 1.$$

If and only if $\frac{7^2}{\frac{-1}{3}} = \frac{(-6)^2}{\frac{1}{7}} + \frac{k^2}{\frac{1}{21}} = (3k - 9)^2$

$$\therefore 2k^2 + 9k + 4 = 0$$
$$\therefore 2k^2 + 8k + k + 4 = 0$$
$$\therefore 2k(k + 4) + 1(k + 4) = 0$$
$$\therefore (k + 4)(2k + 1) = 0$$
$$k = -\ \text{or}\ k = -1/2.$$

$\therefore$ The required tangent planes are $7x - 6y - 4z + 21 = 0$ and $7x - 6y - \frac{z}{2} + \frac{21}{2} = 0$.

i.e., $14x - 12y - z + 21 = 0$.

3) Show that the plane $3x + 12y - 6z - 17 = 0$ touches the conicoid $3x^2 - 6y^2 - 9z^2 + 17 = 0$ and finds the point of contact.

Sol. The equation of the given plane is

$$3x + 12y - 6z - 17 = 0 \qquad (6.8)$$

and the conicoid is

$$3x^2 - 6y^2 + 9z^2 + 17 = 0 \qquad (6.9)$$

Suppose the plane (6.8) touches the conicoid at point (α, β, γ) then the tangent plane at (α, β, γ) is

$$3\alpha x - 6\beta y + 9\gamma z + 17 = 0. \qquad (6.10)$$

Comparing (6.8) and (6.10), we get
$$\frac{3\alpha}{3} = \frac{-6\beta}{12} = \frac{9\gamma}{-6} = \frac{17}{-17}$$
$$\therefore \alpha = -1; \beta = 2; \gamma = \frac{2}{3}.$$

∴ The point of contact is $\left(-1, 2, \frac{2}{3}\right)$ if it satisfies the equation of $3x^2 - 6y^2 + 9z^2 + 17 = 0$.

$$\text{L.H.S.} = 3x^2 - 6y^2 + 9z^2 + 17$$
$$= 3(-1)^2 - 6(2)^2 + 9\left(\frac{2}{3}\right) + 17$$
$$= 0 = \text{R.H.S.}$$

∴ The point $\left(-1, 2, \frac{2}{3}\right)$ is the point of contact of conicoid and plane.

4) Find the equation to the tangent planes to the surface $4x^2 - 5y^2 + 7z^2 + 13 = 0$ parallel to the plane $4x + 20y - 21z = 0$ and also find their points of contact.

Sol. The equation of conicoid is
$$4x^2 - 5y^2 + 7z^2 + 13 = 0, \tag{6.11}$$

and the given plane is
$$4x + 20y - 21z = 0. \tag{6.12}$$

Any plane parallel to the given plane is
$$4x + 20y - 21z + \lambda = 0. \tag{6.13}$$

Let this be the required tangent plane. Let the point of contact be (α, β, γ) then the tangent plane at (α, β, γ) to the conicoid (6.11) is
$$4\alpha x - 5\beta y + 7\gamma z + 13 = 0. \tag{6.14}$$

Comparing (6.13) and (6.14), we get $\frac{4\alpha}{4} = \frac{-5\beta}{20} = \frac{7\gamma}{-21} = \frac{13}{\lambda}$

$$\therefore \alpha = \frac{13}{\lambda}; \beta = \frac{-52}{\lambda}; \gamma = \frac{-39}{\lambda}$$

∴ (α, β, γ) also lies on the conicoid we get $4\alpha^2 - 5\beta^2 + 7\gamma^2 + 13 = 0$

$$\therefore 4\left(\tfrac{13}{\lambda}\right)^2 - 5\left(\tfrac{-52}{\lambda}\right)^2 + 7\left(\tfrac{-39}{\lambda}\right)^2 + 13 = 0$$
$$\therefore 13^2 [4 - 80 + 63] + 13\lambda^2 = 0$$
$$\therefore \lambda^2 = 169$$
$$\therefore \lambda = \pm 13.$$

∴ Point (α, β, γ) is $\left(\frac{13}{\lambda}; \frac{-52}{\lambda}; \frac{-39}{\lambda}\right) = (\pm 1; \mp 4; \mp 3)$.

5) Find the equations to the two planes which contain the line given by $7x + 10y - 30 = 0; 5y - 3z = 0$ and touch the ellipsoid
$$7x^2 + 5y^2 + 3z^2 = 60.$$

Sol. The given equation of surface is
$$7x^2 + 5y^2 + 3z^2 = 60, \tag{6.15}$$

and the line is $7x + 10y - 30 = 0 = 5y - 3z$.
Any plane through this line is $7x + 10y - 30 + \lambda(5y - 3z) = 0$
$$\therefore 7x + 5(2 + \lambda)y - 3\lambda z = 30,$$

which will touch the ellipsoid
$$7x^2 + 5y^2 + 3z^2 = 60$$
$$\therefore \frac{x^2}{\frac{60}{7}} + \frac{y^2}{\frac{60}{5}} + \frac{z^2}{\frac{60}{3}} = 1$$

i.e., $7^2 \left(\frac{60}{7}\right) + 5^2(2+\lambda)^2 \left(\frac{60}{5}\right) + (3\lambda)^2 \left(\frac{60}{3}\right) = (30)^2$

$$\therefore 2\lambda^2 + 5\lambda + 3 = 0$$
$$\therefore 2\lambda^2 + 2\lambda + 3\lambda + 3 = 0$$
$$\therefore (\lambda + 2)(2\lambda + 3) = 0$$
$$\lambda = -1 \text{ or } \lambda = \frac{-3}{2}$$

∴ The required equations are $7x + 5y + 3z = 30$ and $14x + 5y + 9z = 60$.

6.5 Normal to Central Conicoid

The normal at any point of a conicoid is the line through the point perpendicular to the tangent plane at that point to the central conicoid.

Equation of normal at the point to conicoid:
To find the equation of normal at the point (α, β, γ) to the conicoid.
The equation of tangent plane at point (α, β, γ) of the central conicoid
$$ax^2 + by^2 + cz^2 = 1 \tag{6.16}$$

is
$$a\alpha x + b\beta y + c\gamma z = 1. \tag{6.17}$$

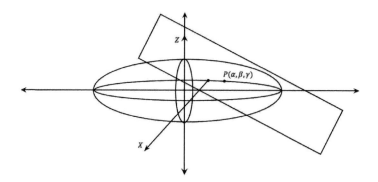

Equation of normal passing through $P(\alpha, \beta, \gamma)$ is

$$\frac{x-\alpha}{a\alpha} = \frac{y-\beta}{b\beta} = \frac{z-\gamma}{c\gamma}$$

where $a\alpha, b\beta, c\gamma$ are the direction ratios of the normal.

Remark:
Let p be the perpendicular distance from the origin to the plane (6.17) is

$$p = \left|\frac{a\alpha(0)+b\beta(0)+c\gamma(0)-1}{\sqrt{(a\alpha)^2+(b\beta)^2+(c\gamma)^2}}\right|$$

$$\therefore p^2 = \frac{1}{(a\alpha)^2+(b\beta)^2+(c\gamma)^2}$$

$$\therefore (a\alpha p)^2 + (b\beta p)^2 + (c\gamma p)^2 = 1.$$

$\therefore$ We can say that $a\alpha p$; $b\beta p$; $c\gamma p$ are the direction cosines of the normal at the point (α, β, γ).

6) Prove that from any given point six normal can be drawn to a central conicoid.

Sol. The equation of the normal to the central conicoid

$$ax^2 + by^2 + cz^2 = 1 \qquad (6.18)$$

at point

(α, β, γ) is $\dfrac{x-\alpha}{a\alpha} = \dfrac{y-\beta}{b\beta} = \dfrac{z-\gamma}{c\gamma}.$ \qquad (6.19)

Let these normal passes through the point (x_1, y_1, z_1)

6.5 Normal to Central Conicoid

∴ Equation (6.19) becomes,

$$\frac{x_1 - \alpha}{a\alpha} = \frac{y_1 - \beta}{b\beta} = \frac{z - \gamma}{c\gamma} = \lambda$$

∴ $x_1 = \alpha(\lambda a + 1)$; ; $y_1 = \beta(\lambda b + 1)$; $z_1 = \gamma(\lambda c + 1)$.

∴ $\alpha = \dfrac{x_1}{\lambda a + 1} ; \beta = \dfrac{y_1}{\lambda b + 1} ; \gamma = \dfrac{z_1}{\lambda c + 1}.$

Since (α, β, γ) is a point on (6.18); so Equation (6.18) becomes;

$$a\alpha^2 + b\beta^2 + c\gamma^2 = 1$$

∴ $a\left(\dfrac{x_1}{\lambda a + 1}\right)^2 + b\left(\dfrac{y_1}{\lambda b + 1}\right)^2 + c\left(\dfrac{z_1}{\lambda c + 1}\right)^2 = 1$

∴ $ax_1^2(\lambda b + 1)^2(\lambda c + 1)^2 + by_1^2(\lambda a + 1)^2(\lambda c + 1)^2 + cz_1^2(\lambda a + 1)^2(\lambda b + 1)^2$
$= (\lambda a + 1)^2(\lambda b + 1)^2(\lambda c + 1)^2$

which is a sixth-degree equation in λ and solving the equation we get six values of λ.

∴ In general, we can draw six normals to the central conicoid passing through points (x_1, y_1, z_1).

A Number of normals from a given point:

Let the ellipsoid be

$$\frac{x^2}{a^2} + \frac{y^2}{b^2} + \frac{z^2}{c^2} = 1. \tag{6.20}$$

Equations of the normal at (x_1, y_1, z_1) are

$$\frac{x - x_1}{\frac{x_1}{a^2}} = \frac{y - y_1}{\frac{y_1}{b^2}} = \frac{z - z_1}{\frac{z_1}{c^2}}.$$

If this passes through (α, β, γ) then

$$\frac{\alpha - x_1}{\frac{x_1}{a^2}} = \frac{\beta - y_1}{\frac{y_1}{b^2}} = \frac{\gamma - z_1}{\frac{z_1}{c^2}} = k$$

∴ $\dfrac{\alpha - x_1}{\frac{x_1}{a^2}} = k \Rightarrow \alpha - x_1 = \dfrac{kx_1}{a^2} \Rightarrow x_1 = \dfrac{\alpha a^2}{a^2 + k}.$

Similarly,

$$y_1 = \frac{\beta b^2}{b^2 + k} \text{ and } z_1 = \frac{\gamma c^2}{c^2 + k} \tag{6.21}$$

194 Central Conicoid

Since (x_1, y_1, z_1) lies on (6.20), we get

$$\frac{x_1^2}{a^2} + \frac{y_1^2}{b^2} + \frac{z_1^2}{c^2} = 1$$

$$\therefore \frac{1}{a^2}\left(\frac{a^2\alpha}{a^2+k}\right)^2 + \frac{1}{b^2}\left(\frac{b^2\beta}{b^2+k}\right)^2 + \frac{1}{c^2}\left(\frac{c^2\gamma}{c^2+k}\right)^2 = 1$$

$$\therefore \frac{a^2\alpha^2}{(a^2+k)^2} + \frac{b^2\beta^2}{(b^2+k)^2} + \frac{c^2\gamma^2}{(c^2+k)^2} = 1$$

$$\therefore a^2\alpha^2(b^2+k)^2(c^2+k)^2 + b^2\beta^2(a^2+k)^2(c^2+k)^2$$
$$+ c^2\gamma^2(a^2+k)^2(b^2+k)^2 = (a^2+k)^2(b^2+k)^2(c^2+k)^2$$

which is a sixth-degree equation in k.

$\therefore$ It gives six values of k corresponding to which six points (x_1, y_1, z_1) are obtained from the Equation (6.21), the normals at which pass through a given point.

Corollary: Foot of normal

By Equation (6.21), $\left[\frac{a^2\alpha}{a^2+k}, \frac{b^2\beta}{b^2+k}, \frac{c^2\gamma}{c^2+k}\right]$ are the coordinates of the foot of normal.

Cubic curve through the feet of six normals from a point:

Let the ellipsoid be

$$\frac{x^2}{a^2} + \frac{y^2}{b^2} + \frac{z^2}{c^2} = 1. \tag{6.22}$$

If the normal at (x_1, y_1, z_1) to the ellipsoid passes through the given point (α, β, γ) then

$$x_1 = \frac{a^2\alpha}{a^2+k}, \quad y_1 = \frac{b^2\beta}{b^2+k}, \quad z_1 = \frac{c^2\gamma}{c^2+k}.$$

The feet of the normals (x_1, y_1, z_1) lie on the curve

$$x = \frac{a^2\alpha}{a^2+k}, \quad y = \frac{b^2\beta}{b^2+k}, \quad z = \frac{c^2\gamma}{c^2+k}. \tag{6.23}$$

where k is a parameter,

To prove that the curve (6.23) is a cubic curve.

To test the degree of the curve we see its intersection with any arbitrary plane.

6.5 Normal to Central Conicoid

The curve (6.23) meets an arbitrary plane

$$ux + vy + wz + d = 0 \qquad (6.24)$$

$$\therefore u\left(\frac{a^2\alpha}{a^2+k}\right) + v\left(\frac{b^2\beta}{b^2+k}\right) + w\left(\frac{c^2\gamma}{c^2+k}\right) + d = 0$$

$$\therefore ua^2\alpha\left(b^2+k\right)\left(c^2+k\right) + vb^2\beta\left(a^2+k\right)\left(c^2+k\right)$$
$$+wc^2\lambda\left(a^2+k\right)\left(b^2+k\right) + d\left(a^2+k\right)\left(b^2+k\right)\left(c^2+k\right) = 0$$

which is a cubic in k, giving three values of k.

$\therefore$ The curve (6.23) is the cubic curve.

Since feet of the normals also lie on the ellipsoid (6.22), we conclude that feet of the six normals from a given point are the six points of intersection of the ellipsoid and a cubic curve.

Quadric cone through six concurrent normals:

To show that the six normals from (α, β, γ) to the ellipsoid lie on a cone of the second degree.

Let the ellipsoid be

$$\frac{x^2}{a^2} + \frac{y^2}{b^2} + \frac{z^2}{c^2} = 1. \qquad (6.25)$$

Now since the normal at (x_1, y_1, z_1) passes through (α, β, γ) we get

$$x_1 = \frac{a^2\alpha}{a^2+k}, \quad y_1 = \frac{b^2\beta}{b^2+k}, \quad z_1 = \frac{c^2\gamma}{c^2+k}.$$

Let the equations of the normal from (α, β, γ) to the ellipsoid be

$$\frac{x-\alpha}{l} = \frac{y-\beta}{m} = \frac{z-\gamma}{n} \qquad (6.26)$$

$$\therefore l = \frac{px_1}{a^2} = \frac{p}{a^2} \cdot \frac{a^2\alpha}{a^2+k}$$

$$\therefore a^2 + k = \frac{p\alpha}{l}a^2 + k = \frac{p\alpha}{l}. \qquad (6.27)$$

Similarly,

$$b^2 + k = \frac{p\beta}{m}, \qquad (6.28)$$

and

$$c^2 + k = \frac{p\gamma}{n}. \qquad (6.29)$$

196 *Central Conicoid*

Multiplying (6.27), (6.28), and (6.29) by $b^2 - c^2, c^2 - a^2, a^2 - b^2$ and adding, we get

$$0 + k(0) = \frac{p\alpha}{l}(b^2 - c^2) + \frac{p\beta}{m}(c^2 - a^2) + \frac{p\gamma}{n}(a^2 - b^2)$$

$$\therefore \frac{\alpha(b^2 - c^2)}{l} + \frac{\beta(c^2 - a^2)}{m} + \frac{\gamma(a^2 - b^2)}{n} = 0. \tag{6.30}$$

Eliminating l, m, n from (6.26) and (6.30), the locus of the normals (6.26) is

$$\frac{\alpha(b^2 - c^2)}{x - \alpha} + \frac{\beta(c^2 - a^2)}{y - \beta} + \frac{\gamma(a^2 - b^2)}{z - \gamma} = 0$$

$$\therefore \alpha(b^2 - c^2)(y - \beta)(z - \gamma) + \beta(c^2 - a^2)(z - \gamma)(x - \alpha)$$
$$+ \gamma(a^2 - b^2)(x - \alpha)(y - \beta) = 0$$

which is a cone of the second-degree.

10) If $P, Q, R;; P^1, Q^1, R^1$ are the feet of six normals from a point to the ellipsoid $\frac{x^2}{a^2} + \frac{y^2}{b^2} + \frac{z^2}{c^2} = 1$, and the plane PQR is given by $lx + my + nz = p$; then the plane $P^1 Q^1 R^1$ is given by $\frac{x}{a^2 l} + \frac{y}{b^2 m} + \frac{z}{c^2 n} + \frac{1}{p} = 0$.

Sol. The equation of the ellipsoid is

$$\frac{x^2}{a^2} + \frac{y^2}{b^2} + \frac{z^2}{c^2} - 1 = 0, \tag{6.31}$$

and the plane PQR is

$$lx + my + nz - p = 0. \tag{6.32}$$

Let the required equation of the plane $P^1 Q^1 R^1$ be

$$l^1 x + m^1 y + n^1 z - p^1 = 0. \tag{6.33}$$

The joint equation of the planes PQR and $P^1 Q^1 R^1$ is

$$(lx + my + nz - p)(l^1 x + m^1 y + n^1 z - p^1) = 0. \tag{6.34}$$

$\therefore$ Equation of conicoid through the points of intersection of the ellipsoid (6.31) and pair of planes (6.34) is

$$\left(\frac{x^2}{a^2} + \frac{y^2}{b^2} + \frac{z^2}{c^2} - 1\right) + k(lx + my + nz - p)(l^1 x + m^1 y + n^1 z - p^1) = 0. \tag{6.35}$$

If it is the same as the equation of the cone through the feet $P, Q, R;; P^1, Q^1, R^1$ of the six normals from the given point to the ellipsoid, then

Coefficient of $x^2 = 0 \Rightarrow \frac{1}{a^2} + kll^1 = 0 \Rightarrow l^1 = \frac{-1}{kla^2}$

Coefficient of $y^2 = 0 \Rightarrow \frac{1}{b^2} + kmm^1 = 0 \Rightarrow m^1 = \frac{-1}{kmb^2}$

Coefficient of $z^2 = 0 \Rightarrow \frac{1}{c^2} + knn^1 = 0 \Rightarrow n^1 = \frac{-1}{knc^2}$

Constant them $= 0 \Rightarrow -1 + kpp^1 = 0 \Rightarrow p^1 = \frac{1}{kp}$.

Substituting these values in (6.33), we get

$$\frac{-x}{kla^2} - \frac{y}{kmb^2} - \frac{z}{knc^2} - \frac{1}{kp} = 0$$

$$\therefore \frac{x}{la^2} + \frac{y}{mb^2} + \frac{z}{nc^2} + \frac{1}{p} = 0$$

is the required equation of the plane $P^1 Q^1 R^1$.

6.6 Plane of Contact

To find the equation of plane of contact of the point (x_1, y_1, z_1) with respect to conicoid $ax^2 + by^2 + cz^2 = 1$.

Let (x^1, y^1, z^1) be the point of contact any tangent plane to the conicoid

$$ax^2 + by^2 + cz^2 = 1. \tag{6.36}$$

$\therefore$ Tangent plane at (x^1, y^1, z^1) to (6.36) is $axx^1 + byy^1 + czz^1 = 1$,
$\therefore$ locus of the points of contact (x^1, y^1, z^1) is $ax_1 x + by_1 y + cz_1 z = 1$
which is the required plane of contact.

6.7 Polar Plane of a Point

The given conicoid is
$$ax^2 + by^2 + cz^2 = 1. \tag{6.37}$$

Let $P(x_1, y_1, z_1)$ be the given point and PQR be any line through P which meets (6.37) in Q and R.
Let $S(x, y, z)$ be the harmonic conjugate of P with respect to Q and R.
Let Q divide PS in the ratio $k : 1$.
Coordinates of Q are $\left(\frac{kx+x_1}{k+1}, \frac{ky+y_1}{k+1}, \frac{kz+z_1}{z+1} \right)$.

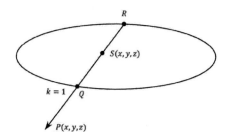

Since Q lies on the conicoid CD.

$$\therefore a\left(\frac{kx+x_1}{k+1}\right)^2 + b\left(\frac{ky+y_1}{k+1}\right)^2 + c\left(\frac{kz+z_1}{k+1}\right)^2 = 1$$

$$\therefore a(kx+x_1)^2 + b(ky+y_1)^2 + c(kz+z_1)^2 = (k+1)^2$$

$$\therefore k^2\left(ax^2+by^2+cz^2-1\right) + 2k\left(axx_1+byy_1+czz_1-1\right) + \left(ax_1^2+by_1^2+cz_1^2-1\right) = 0 \quad (6.38)$$

which is a quadratic equation in k.

Since PS is divided harmonically, i.e., internally and externally in that same ratio at Q and R.

$\therefore$ The quadratic equation (6.38) has equal and opposite roots.
$\therefore$ Sum of roots $= 0$.
i.e., Coefficient of $k = 0$.

$$\therefore axx_1 + byy_1 + czz_1 - 1 = 0$$

$$\therefore axx_1 + byy_1 + czz_1 = 1$$

which is the equation of the required polar plane of P.

Corollary:

If P lies on the corticoid, the polar plane at P becomes the tangent plane at P.

11) A tangent plane to the ellipsoid $\frac{x^2}{a^2} + \frac{y^2}{b^2} + \frac{z^2}{c^2} = 1$ meets the coordinate axes in A, B, and C. Find the bonus of the centroid of the $\triangle ABC$.

Sol. Let $P(x_1, y_1, z_1)$ be any point on the ellipsoid

$$\frac{x^2}{a^2} + \frac{y^2}{b^2} + \frac{z^2}{c^2} = 1, \quad (6.39)$$

$$\therefore \frac{x_1^2}{a^2} + \frac{y_1^2}{b^2} + \frac{z_1^2}{c^2} = 1. \qquad (6.40)$$

Equation of tangent plane at $P(x_1, y_1, z_1)$ to (6.39) is

$$\frac{xx_1}{a^2} + \frac{yy_1}{b^2} + \frac{zz_1}{c^2} = 1. \qquad (6.41)$$

This meets $x\ axis\ (y = 0, z = 0)$, where $\frac{xx_1}{a^2} = 1 \Rightarrow x = \frac{a^2}{x_1}$.

$\therefore$ (6.41) meets $x\ axis$ in the point $A\left(\frac{a^2}{x_1}, 0, 0\right)$.

Similarly, it meets $y\ axis$ in $B\left(0, \frac{b^2}{y_1}, 0\right)$, and $z\ axis$ in $C\left(0, 0, \frac{c^2}{z_1}\right)$.

If $G(\alpha, \beta, \gamma)$ be the centroid of $\triangle ABC$;

$$\alpha = \frac{\frac{a^2}{x_1} + 0 + 0}{3} = \frac{a^2}{3x_1}.$$

Similarly, $\beta = \frac{b^2}{3y_1}$ and $\gamma = \frac{c^2}{3z_1}$.

$$\therefore x_1 = \frac{a^2}{3\alpha},\ y_1 = \frac{b^2}{3\beta},\ z_1 = \frac{c^2}{3\gamma}.$$

Substituting the value of x_1, y_1, z_1 in (6.40), we get

$$\frac{1}{a^2}\frac{a^4}{9\alpha^2} + \frac{1}{b^2}\frac{b^4}{9\beta^2} + \frac{1}{c^2}\frac{c^4}{9\gamma^2} = 1$$

$$\therefore \frac{a^2}{\alpha^2} + \frac{b^2}{\beta^2} + \frac{c^2}{\gamma^2} = 9$$

$\therefore$ Locus of $G(\alpha, \beta, \gamma)$ is $\frac{a^2}{x^2} + \frac{b^2}{y^2} + \frac{c^2}{z^2} = 9$.

12) Find the condition that the line $\frac{x-2}{l} = \frac{y-1}{m} = \frac{z-3}{n}$ may touch the ellipsoid $3x^2 + 8y^2 + z^2 = c^2$.

Sol. The given equation of the line is $\frac{x-2}{l} = \frac{y-1}{m} = \frac{z-3}{n}$
and of the ellipsoid is $3x^2 + 8y^2 + z^2 = c^2$.

$$\therefore \frac{3}{c^2}x^2 + \frac{8}{c^2}y^2 + \frac{z^2}{c^2} = 1.$$

The condition for becoming a tangent line is $al\alpha + bm\beta + cn\gamma = 0$.
i.e., $\frac{3}{c^2}(2)l + \frac{8}{c^2}(1)m + \frac{1}{c^2}(3)n = 0$.

$\therefore 6l + 8m + 3n = 0$ is the required condition.

13) The normal at a point P of the ellipsoid $\frac{x^2}{a^2} + \frac{y^2}{b^2} + \frac{z^2}{c^2} = 1$ meets the principal planes in G_1, G_2, G_3; (i) Show that $PG_1 : PG_2 : PG_3 = a^2 : b^2 : c^2$ (ii) If $PG_1^2 + PG_2^2 + PG_3^2 = k^2$, find the locus of P.

Sol. The given equation of ellipsoid is

$$\frac{x^2}{a^2} + \frac{y^2}{b^2} + \frac{z^2}{c^2} = 1. \tag{6.42}$$

Let $P(x_1, y_1, z_1)$ be any point.

The equations of the normal at $P(x_1, y_1, z_1)$ to (6.42) in the actual direction cosines form are $\frac{x-x_1}{\frac{px_1}{a^2}} = \frac{y-y_1}{\frac{py_1}{b^2}} = \frac{z-z_1}{\frac{pz_1}{c^2}} = t$.

$\therefore$ Any point on this normal is $\left(x_1 + \frac{tpx_1}{a^2}, y_1 + \frac{tpy_1}{b^2}, z_1 + \frac{tpz_1}{c^2}\right)$.

If it lies on the YZ plane i.e., $x = 0$, then

$$x_1 + \frac{px_1 r}{a^2} = 0 \Rightarrow r = \frac{-a^2}{p}$$

i.e., $PG_2 = \frac{-b^2}{p}$ and $PG_3 = \frac{-c^2}{p}$

(i) $\therefore PG_1 : PG_2 : PG_3 = \frac{-a^2}{p} : \frac{-b^2}{p} : \frac{-c^2}{p} = a^2 : b^2 : c^2$

(ii) Given $PG_1^2 + PG_2^2 + PG_3^2 = k^2$

$$\therefore \frac{a^4}{p^2} + \frac{b^4}{p^2} + \frac{c^4}{p^2} = k^2$$

$$\therefore \frac{1}{p^2} = \frac{k^2}{a^4 + b^4 + c^4}. \tag{6.43}$$

But P is the perpendicular distance from $(0, 0, 0)$ on the tangent plane

$\therefore \frac{xx_1}{a^2} + \frac{yy_1}{b^2} + \frac{zz_1}{c^2} = 1$ at (x_1, y_1, z_1) to (6.42)

$$= \frac{-1}{\sqrt{\frac{x_1^2}{a^4} + \frac{y_1^2}{b^4} + \frac{z_1^2}{c^4}}}.$$

By (6.40), $\frac{1}{p^2} = \frac{x_1^2}{a^4} + \frac{y_1^2}{b^4} + \frac{z_1^2}{c^4} = \frac{k^2}{a^4+b^4+c^4}$

$\therefore$ Locus of $P(x_1, y_1, z_1)$ is

$$\frac{x^2}{a^4} + \frac{y^2}{b^4} + \frac{z^2}{c^4} = \frac{k^2}{a^4 + b^4 + c^4}. \tag{6.44}$$

Also, P lies on (6.42). Thus P lies on the curve of the intersection of the two ellipsoids (6.42) and (6.43).

14) Prove that for all values of λ, the normals to the conicoid

$$\frac{x^2}{a^2 + \lambda} + \frac{y^2}{b^2 + \lambda} + \frac{z^2}{c^2 + \lambda} = 1,$$

which pass through a given point (α, β, γ) meet the plane $z = 0$ in points on the conic $(b^2 - c^2)\beta x + (c^2 - a^2)\alpha y + (a^2 - b^2) xy = 0, z = 0.$

Sol. The equation of the quadric cone containing the normals to

$$\frac{x^2}{a^2 + \lambda} + \frac{y^2}{b^2 + \lambda} + \frac{z^2}{c^2 + \lambda} = 1,$$

drawn from the point (α, β, γ) is

$$\sum \frac{1}{a^2 + \lambda} \alpha \left(\frac{1}{b^2 + \lambda} - \frac{1}{c^2 + \lambda} \right) \frac{1}{x - \alpha} = 0$$

$$\therefore \sum \frac{\alpha (c^2 - b^2)}{x - \alpha} = 0.$$

Thus, it meets the plane $z = 0$, where

$$\frac{\alpha (c^2 - b^2)}{x - \alpha} + \frac{\beta (a^2 - c^2)}{y - \beta} - (b^2 - a^2) = 0$$

$$\therefore \alpha(y - \beta)(c^2 - b^2) + \beta(x - \alpha)(a^2 - c^2) - (x - \alpha)(y - \beta)(b^2 - a^2) = 0$$

$$\therefore (b^2 - c^2)\beta x + (c^2 - a^2)\alpha y + (a^2 - \beta)xy = 0.$$

Exercise:

1) Show that the plane $x + 2y + 3z = 2$ touches the conicoid $x^2 - 2y^2 + 3z^2 = 2$.

2) Obtain the tangent planes to the ellipsoid $\frac{x^2}{a^2} + \frac{y^2}{b^2} + \frac{z^2}{c^2} = 1$, which are parallel to the plane $lx + my + nz = 0$. If $2r$ is the distance between two parallel tangent planes to the ellipsoid, prove the line through the origin and perpendicular to the planes lies on the cone $x^2 (a^2 - r^2) + y^2 (b^2 - r^2) + z^2 (c^2 - r^2) = 0.$

3) Find the equation to the tangent planes to $2x^2 - 6y^2 + 3z^2 = 5$ which passes through the line $x + 9y - 3z = 0 = 3x - 3y + 6z - 5$.

Answer: $2x - 12y + 9z = 5, 4x + 6y + 3z = 5$

4) Find the equation of the tangent plane to the surface $3x^2 - 6y^2 + 9z^2 + 17 = 0$ parallel to the plane $x + 4y - 2z = 0$.

Answer: $3x + 12 - 6z = \pm 17$

5) Prove that two normals to the ellipsoid $\frac{x^2}{a^2} + \frac{y^2}{b^2} + \frac{z^2}{c^2} - 1$, lie in the plane $lx + my + nz = 0$, and the line joining their feet has direction cosines proportional to $a^2(b^2 - c^2)mn$, $b^2(c^2 - a^2)nl$, $c^2(a^2 - b^2)lm$. Also, obtain the coordinates of these points.

7

Miscellaneous Examples using MATLAB

1) Using MATLAB plot the following plane: 2x+3y+4z+10=0.

Sol.
   A=2;
   B=3;
   C=4;
   D=10;

```
[x y] = meshgrid(-1:0.1:1);   % Generate x and y data
z = -1/C*(A*x + B*y + D);    % Solve for z data
surf(x,y,z)  %Plot the surface
```

Output:

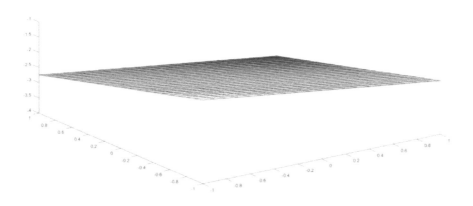

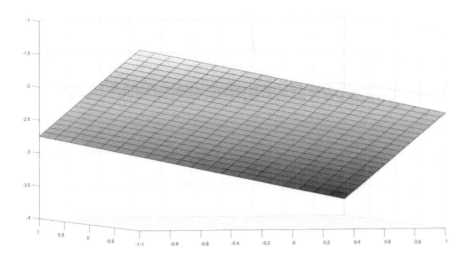

2) Finds the equation of a plane passing through three points $A(1, 2, 3)$, $B(-1, 2, -3)$, and $C(1, 2, -3)$ in three-dimensional space.

Sol. To find the equation of a plane passing through three points $A(1, 2, 3)$, $B(-1, 2, -3)$, and $C(1, 2, -3)$ first we create a function file named as Plane_3Points.m and the code is given as follow: 2 then we run the program

```
function [a,b,c,d]=Plane_3Points(A,B,C)
a=(B(2)-A(2))*(C(3)-A(3))-(C(2)-A(2))*(B(3)-A(3));
b=(B(3)-A(3))*(C(1)-A(1))-(C(3)-A(3))*(B(1)-A(1));
c=(B(1)-A(1))*(C(2)-A(2))-(C(1)-A(1))*(B(2)-A(2));
d=-(a*A(1)+b*A(2)+c*A(3));
end
```

A= [1 2 3];
B= [−1 2 −3];
C= [1 2 −3];

[a,b,c,d]=Plane_3Points (A,B,C)

Output:
a =
    0

$b = $
$\quad -12$
$c = $
$\quad 0$
$d = $
$\quad 24$

∴ The equation of the plane is $-12y + 24 = 0$.

3) Finds the equation of a plane passing through three points $A(1, 0, 3)$, $B(0, 2, 3)$, and $C(-2, 4, -3)$ in three-dimensional space.

Sol.

```
function [a,b,c,d]=Plane_3Points(A,B,C)
a=(B(2)-A(2))*(C(3)-A(3))-(C(2)-A(2))*(B(3)-A(3));
b=(B(3)-A(3))*(C(1)-A(1))-(C(3)-A(3))*(B(1)-A(1));
c=(B(1)-A(1))*(C(2)-A(2))-(C(1)-A(1))*(B(2)-A(2));
d=-(a*A(1)+b*A(2)+c*A(3));
end
```

A= [1 0 3];
B= [0 2 3];
C= [−2 4 −3];
[a,b,c,d]=Plane_3Points (A,B,C)

Output:

$a = $
$\quad -12$
$b = $
$\quad -6$
$c = $
$\quad 2$
$d = $
$\quad 6$

∴ The equation of the plane is $-12x - 6y + 2z + 6 = 0$

4) Plot a plane based on a normal vector and a point in Matlab.

Sol.

```
point = [1,2,3];
normal = [1,2,2];
t=(0:10:360)';
circle0=[cosd(t) sind(t) zeros(length(t),1)];
r=vrrotvec2mat(vrrotvec([0 0 1],normal));
circle=circle0*r'+repmat(point,length(circle0),1);
patch(circle(:,1),circle(:,2),circle(:,3),.5);
axis square; grid on;
%add line
line=[point;point+normr(normal)]
hold
on;plot3(line(:,1),line(:,2),line(:,3),'LineWidth'
,3)
```

Output:

Line=

1.0000   2.0000   3.0000
7.3333   2.6667   3.6667

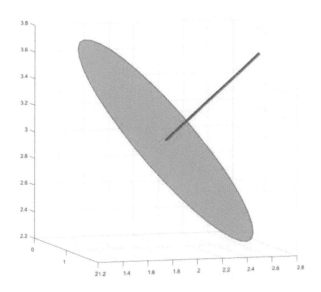

5) Find the angle between the planes $2x - y + 2z = 3$ and $3x + 6y + 2z = 4$.
Sol.

First we assign $2x - y + 2z = 3$ as a vector1 and $3x + 6y + 2z = 4$ as a vector2 and then we apply the formula to find angle between two planes.
vector1 = $[2,-1,2]$;
vector2 = $[3,6,2]$;
Theta = atan2d(norm(cross(vector1, vector2)), dot(vector1, vector2))

Output:

Theta = 79.0194

6) Is the origin in the acute or obtuse angle between the planes $x + y - z = 3$ and $x - 2y + z = 3$.
Sol.

```
%Enter coefficients for 1st plane
a1=input('enter the coefficient of x for the 1st
plane :');
b1=input('enter the coefficient of y for the 1st
plane :');
c1=input('enter the coefficient of z for the 1st
plane :');
%Enter coefficients for 2nd plane
a2=input('enter the coefficient of x for the 2nd
plane :');
b2=input('enter the coefficient of y for the 2nd
plane :');
c2=input('enter the coefficient of z for the 2nd
plane :');

d=(a1*a2)+(b1*b2)+(c1*c2);
fprintf('(a1*a2)+(b1*b2)+(c1*c2) = %d \n',d);
% The origin in the acute or obtuse angle between
the planes
if d>0
fprintf('The origin lies in the obtuse angle
between the planes. ')
else
fprintf('The origin lies in the acute angle
between the planes. ')
end
```

Output:
(a1*a2)+(b1*b2)+(c1*c2) = −2
The origin lies in the acute angle between the planes.

7) Find the angle between the lines
$$3x + 2y + z - 5 = 0 = x + y - 2z - 3$$
$$2x - y - z = 0 = 7x + 10y - 8z.$$

Sol. First, we find the direction ratios of the given lines

$$\left.\begin{array}{r}3x + 2y + z = 5 \\ x + y - 2z = 3\end{array}\right\} \quad (7.1)$$

and

$$\left.\begin{array}{r}2x - y - z = 0 \\ 7x + 10y - 8z = 0\end{array}\right\} \quad (7.2)$$

Let l_1, m_1, n_1 be the direction ratios of the line (7.1) then

$$\frac{l_1}{(2)(-2) - (1)(1)} = \frac{m_1}{(1)(1) - (3)(-2)} = \frac{n_1}{3(1) - (2)(1)}$$

$$\therefore \frac{l_1}{-5} = \frac{m_1}{7} = \frac{n_1}{1}$$

∴ The direction ratios of the line (7.1) are −5, 7, 1
Let l_2, m_2, n_2 be the direction ratios of the line (7.2) then

$$\frac{l_2}{(-1)(-8) - (-1)(10)} = \frac{m_2}{(-1)(7) - (-8)(2)} = \frac{n_2}{(2)(10) - (-1)(7)}$$

$$\therefore \frac{l_2}{18} = \frac{m_2}{9} = \frac{n_2}{27}$$

$$\therefore \frac{l_2}{2} = \frac{m_2}{1} = \frac{n_2}{3}$$

∴ The direction ratios of the line (7.2) are 2, 1, 3.

Input:

```
%Enter coefficients for 1st line
a1=input('enter the direction cosine l1 for the
1st line :');
b1=input('enter the direction cosine m1 for the
1st line :');
c1=input('enter the direction cosine n1 for the
1st line :');
%Enter coefficients for 2nd line
a2=input('enter the direction cosine l2 for the
2nd line :');
b2=input('enter the direction cosine m2 for the
2nd line :');
c2=input('enter the direction cosine n2 for the
2nd line :');

d1=(a1*a2)+(b1*b2)+(c1*c2);
d2=sqrt((a1*a1)+(b1*b1)+(c1*c1));
d3=sqrt((a2*a2)+(b2*b2)+(c2*c2));
D1=d2*d3;
D=acosd(d1/D1);
fprintf('The angle between the line is %d',D);
```

Output:
The angle between the lines is 90 degrees.

8) Find the angle between the line $\frac{x+1}{2} = \frac{y}{3} = \frac{z-3}{6}$ and the plane $3x + y + z = 7$.

Sol.

```
% To find the angle between a line and a plane
a=input('enter the coefficient of x for a plane
:');
b=input('enter the coefficient of y for a plane
:');
c=input('enter the coefficient of z for a plane
:');
%Enter coefficients for 2nd line
```

```
l=input('enter the direction cosine l for a line
:');
m=input('enter the direction cosine m for a line
:');
n=input('enter the direction cosine n for a line
:');

if (a*l+b*m+c*n==0)

    fprintf('The line is parallel to the plane and
angle is 0.\n ')

elseif (l/a==m/b==n/c)
    fprintf('The line is perpendicular to the
plane.\n')

else

d1=(a*l)+(b*m)+(c*n);
d2=sqrt((l*l)+(m*m)+(n*n));
d3=sqrt((a*a)+(b*b)+(c*c));
D1=d2*d3;
D=asind(d1/D1);
fprintf('The angle between the line is %f in
degree.\n' ,D);

end
```

Input 1:
enter the coefficient of x for a plane :3
enter the coefficient of y for a plane :1
enter the coefficient of z for a plane :1
enter the direction cosine l for a line :2
enter the direction cosine m for a line :3
enter the direction cosine n for a line :6

Output 1:
The angle between the lines is 40.247880 degrees.

Input 2:
enter the coefficient of x for a plane :1 8 enter the coefficient of y for a plane :1

enter the coefficient of z for a plane :1
enter the direction cosine l for a line :1
enter the direction cosine m for a line :1
enter the direction cosine n for a line :1

Output 2:

The line is perpendicular to the plane.

Input 3:

enter the coefficient of x for a plane :1
enter the coefficient of y for a plane :−1
enter the coefficient of z for a plane :0
enter the direction cosine l for a line :1
enter the direction cosine m for a line :1
enter the direction cosine n for a line :1

Output 3:

The line is parallel to the plane and the angle is 0.

9) Examine the nature of the intersection of the following sets of planes:

(i) $2x + 3y - z - 2 = 0$, $3x + 3y + z - 4 = 0$, $x - y + 2z - 5 = 0$

(ii) $4x - 5y - 2z - 2 = 0$, $5x - 4y + 2z + 2 = 0$, $2x + 2y + 8z - 1 = 0$

Sol.

```
% To find whether the three plane to meet in a
point:
%Enter coefficients for 1st plane
a1=input('enter the coefficient of x for the 1st
plane :');
b1=input('enter the coefficient of y for the 1st
plane :');
c1=input('enter the coefficient of z for the 1st
plane :');
d1=input('enter the value of constant for the 1st
plane :');
%Enter coefficients for 2nd plane
```

```
a2=input('enter the coefficient of x for the 2nd
plane :');
b2=input('enter the coefficient of y for the 2nd
plane :');
c2=input('enter the coefficient of z for the 2nd
plane :');
d2=input('enter the value of constant for the 2nd
plane :');
%Enter coefficients for 3rd plane
a3=input('enter the coefficient of x for the 3rd
plane :');
b3=input('enter the coefficient of y for the 3rd
plane :');
c3=input('enter the coefficient of z for the 3rd
plane :');
d3=input('enter the value of constant for the 3rd
plane :');

D4=[a1 b1 c1;a2 b2 c2;a3 b3 c3]

D3=[a1 b1 d1;a2 b2 d2;a3 b3 d3]

D1=det(D4)
D2=det(D3)

if (D1~=0)

    fprintf('The given three planes intersect at a
point. \n ');

elseif (D2~=0)
    fprintf('The three planes form a triangular
prism. \n');

else

fprintf('The three planes intersect in a line.
\n');

end
```

Input 1:

enter the coefficient of x for the 1st plane :2
enter the coefficient of y for the 1st plane :3
enter the coefficient of z for the 1st plane :−1
enter the value of constant for the 1st plane :−2
enter the coefficient of x for the 2nd plane :3
enter the coefficient of y for the 2nd plane :3
enter the coefficient of z for the 2nd plane :1
enter the value of constant for the 2nd plane :−4
enter the coefficient of x for the 3rd plane :1
enter the coefficient of y for the 3rd plane :−1
enter the coefficient of z for the 3rd plane :2
enter the value of constant for the 3rd plane :−5

Output 1:

D4 =

2	3	-1
3	3	1
1	-1	2

D3 =

2	3	-2
3	3	-4
1	-1	-5

D1 =

   5.0000

D2 =

   7.0000

The given three planes intersect at a point.

Input 2:

enter the coefficient of x for the 1st plane :4
enter the coefficient of y for the 1st plane :−5
enter the coefficient of z for the 1st plane :−2
enter the value of constant for the 1st plane :−2
enter the coefficient of x for the 2nd plane :5
enter the coefficient of y for the 2nd plane :−4
enter the coefficient of z for the 2nd plane :2
enter the value of constant for the 2nd plane :2
enter the coefficient of x for the 3rd plane :2
enter the coefficient of y for the 3rd plane :2
enter the coefficient of z for the 3rd plane :8
enter the value of constant for the 3rd plane :−1

Output 2:

D4 =

```
   4   -5   -2
   5   -4    2
   2    2    8
```

D3 =

```
   4   -5   -2
   5   -4    2
   2    2   -1
```

D1 =

   0

D2 =

   -81

The three planes form a triangular prism.

10) Draw a sphere with a radius of 4 with a center (5,-5,3).

Sol.

```
[X,Y,Z] = sphere;
r = input('Enter radius of a sphere : ');
a = input('Enter x-coordinate of the center: ');
b = input('Enter y-coordinate of the center: ');
c = input('Enter z-coordinate of the center: ');
surf(X+a,Y+b,Z+c)
axis equal
```

Input:

Enter radius of a sphere: 4
Enter x-coordinate of the center: 5
Enter y-coordinate of the center: -5
Enter z-coordinate of the center: 3

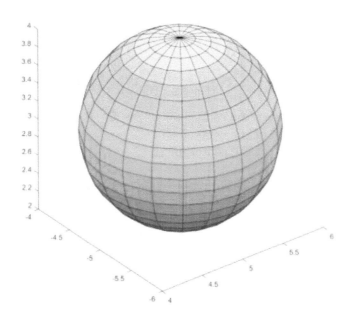

Output:

11) Draw a sphere with a radius of 4 by multiplying the coordinates of the unit sphere. Plot the second sphere with center $(5, -5, 3)$.

Sol.

```
[X,Y,Z] = sphere;
r = input('Enter how many times radius required a new sphere than a unit vector: ');
a = input('Enter x-coordinate of the center: ');
b = input('Enter y-coordinate of the center: ');
c = input('Enter z-coordinate of the center: ');
surf(X,Y,Z)
axis equal

hold on

X2 = X * r;
Y2 = Y * r;
Z2 = Z * r;

surf(X2+a,Y2+b,Z2+c)
```

Input:

Enter how many times the radius required a new sphere than a unit vector: 4
Enter the x-coordinate of the center: 5
Enter the y-coordinate of the center: −5
Enter the z-coordinate of the center: 0

Output:

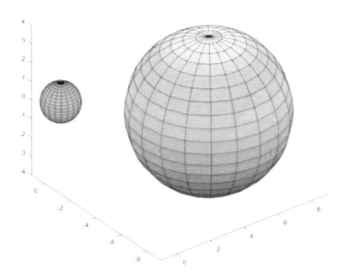

12) Draw a simple cone.

Sol.

r = linspace(0,2*pi) ;
th = linspace(0,2*pi) ;
[R,T] = meshgrid(r,th) ;
X = R.*cos(T) ;
Y = R.*sin(T) ;
Z = R ;
surf(X,Y,Z)

Output:

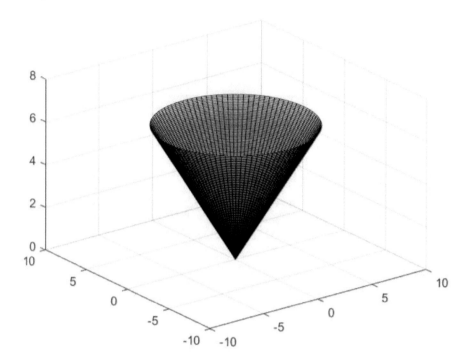

13) Draw a right circular cone.

Sol.
r = linspace(0,−2*pi) ;
th = linspace(0,-2*pi) ;
[R,T] = meshgrid(r,th) ;
X = R.*cos(T) ;
Y = R.*sin(T) ;
Z = R ;
surf(X,Y,Z)

Output:

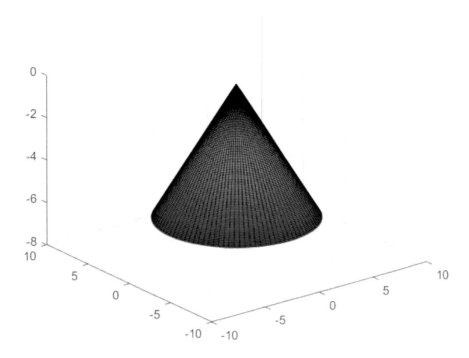

14) Draw a right circular cone whose base is on the positive y-axis.

Sol.

r = linspace(0,2*pi) ;
th = linspace(0,2*pi);
[R,T] = meshgrid(r,th);
X = R.*cos(T) ;
Y = R ;
Z = R.*sin(T) ;
surf(X,Y,Z)

Output:

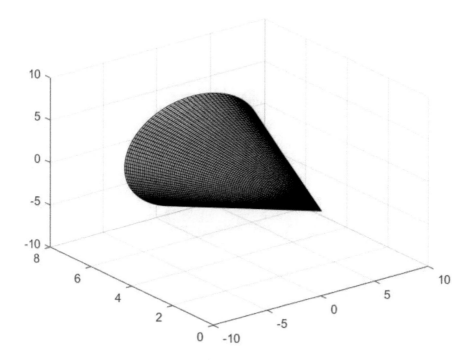

15) Draw a right circular cone whose base is on the negative y-axis.

Sol.
r = linspace(0,2*pi) ;
th = linspace(0,2*pi) ;
[R,T] = meshgrid(r,th) ;
X = R.*cos(T) ;
Y =-R ;
Z = R.*sin(T) ;
surf(X,Y,Z)

Output:

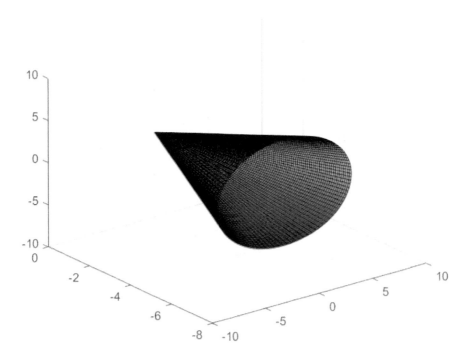

16) Draw a cylinder of $x^2 + y^2 = 1$ in three-dimensional space.
Sol.

```
close all

% initialize the grid
z=-5:5;
theta=linspace(0,2*pi,40);
[z,theta]=meshgrid(y,theta);

% calculate x and z
x=cos(theta);
y=sin(theta);

% plot the mesh
mesh(x,y,z)

% grid and box
grid on
box on

% axis equal
axis equal

% adjust the view
view([200,20])

% annotate the plot
xlabel('x-axis','Fontname','Times','Fontsize',10)
ylabel('y-axis','Fontname','Times','Fontsize',10)
zlabel('z-axis','Fontname','Times','Fontsize',10)
title('The graph of x^2 + y^2 = 1 in three-space.','Fontname','Times','Fontsize',10)

% white background
set(gcf,'color','white')

% save the plot
set(gca,'Fontname','Times','Fontsize',10)
set(gcf,'PaperPosition',[0,0,4/4*3,3/4*3])
```

Output:

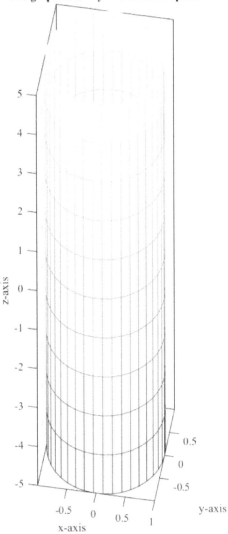

The graph of $x^2 + y^2 = 1$ in three-space.

17) Draw a cylinder of $x^2 + z^2 = 1$ in three-dimensional space.

Sol.

```
% close all open figure windows
close all

% initialize the grid
y=-2:4;
theta=linspace(0,2*pi,40);
[y,theta]=meshgrid(y,theta);

% calculate x and z
x=cos(theta);
z=sin(theta);

% plot the mesh
mesh(x,y,z)

% grid and box
grid on
box on

% axis equal
axis equal

% adjust the view
view([50,10])

% annotate the plot
xlabel('x-axis','Fontname','Times','Fontsize',10)
ylabel('y-axis','Fontname','Times','Fontsize',10)
zlabel('z-axis','Fontname','Times','Fontsize',10)
title('The graph of x^2 + z^2 = 1 in three-space.','Fontname','Times','Fontsize',10)

% white background
set(gcf,'color','white')

% save the plot
set(gca,'Fontname','Times','Fontsize',10)
set(gcf,'PaperPosition',[0,0,4/4*3,3/4*3])
```

Output:

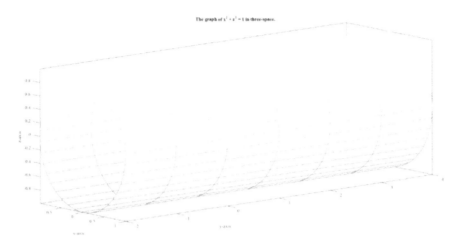

18) Draw a cylinder of $y^2 + z^2 = 1$ in three-dimensional space.

Sol.
```
% close all open figure windows
close all

% initialize the grid
x=-2:4;
theta=linspace(0,2*pi,40);
[x,theta]=meshgrid(x,theta);

% calculate x and z
y=cos(theta);
z=sin(theta);

% plot the mesh
mesh(x,y,z)

% grid and box
grid on
box on

% axis equal
axis equal
```

```
% adjust the view
view([50,10])

% annotate the plot
xlabel('x-axis','Fontname','Times','Fontsize',10)
ylabel('y-axis','Fontname','Times','Fontsize',10)
zlabel('z-axis','Fontname','Times','Fontsize',10)
title('The graph of x^2 + z^2 = 1 in three-space.','Fontname','Times','Fontsize',10)

% white background
set(gcf,'color','white')

% save the plot
set(gca,'Fontname','Times','Fontsize',10)
set(gcf,'PaperPosition',[0,0,4/4*3,3/4*3])
```

Output:

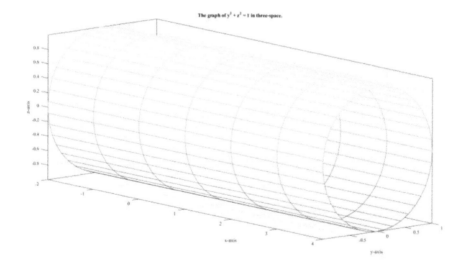

Index

A
Arbitrary 2, 17, 194
Axes 4, 23, 45, 72, 104, 147, 198

C
Centroid 8, 104, 198, 199
Circumcenter 14, 15
Conditions 15, 55, 82, 152
Constant 3, 29, 42, 102, 156, 214
Coordinate 3, 13, 35, 45, 72, 104, 147, 185, 198, 215
Coordinate planes 5, 35, 43
Coordinates 3, 7, 24, 43, 73, 102, 185, 216
Coplanar 10, 13, 56, 58, 93
Cosines 2, 4, 26, 45, 76, 134, 178, 202
Cross multiplication 12, 14, 20

D
Direction cosines 2–4, 45, 53, 134, 155, 178, 200
Distance 9, 26, 44, 69, 71, 88, 105, 122, 161

E
Externally 5, 25, 125, 198

F
Fixed 8, 45, 50, 89, 132, 138, 156, 175
Fixed point 8, 45, 50, 130, 138, 156

I
Intercept 6, 7, 8, 89, 130
Intercept form 6, 7, 8
Internally 5, 24, 132, 198
Intersection 9, 16, 17, 19, 42, 45, 74, 80, 106, 156, 201
Intersects 12, 43, 115, 133, 169, 185

L
Locus 1, 39, 102, 141, 200

N
Negative 4, 25, 105, 220
Normal 2, 3, 4, 116, 191, 200, 200
Normal form 2, 3, 4, 27

O
Origin 2, 23, 89, 102, 137, 150, 178, 208

P
Parallel 3, 6, 78, 122, 167, 200

Passing 3, 5, 95, 124, 192
Perpendicular 2, 22, 78, 105, 122, 154, 180, 211
Perpendiculars 13, 95, 153, 164
Plane 1, 5, 67, 107, 175, 201, 204
Planes 5, 31, 83, 190, 201, 207
Points 1, 26, 98, 161
Positive 3, 23, 28
Projections 2, 35, 37, 45
Proportional 4, 41, 53, 155, 160, 202

R
Ratios 4, 6, 21, 52, 148, 153, 161, 208
Reciprocals 8

Rectangular 4, 73, 83, 140, 141

S
Straight line 2, 35, 45, 47, 56, 133, 138, 175, 183

T
Triangle 8, 14, 35, 44, 130, 176

V
Value 2, 18, 87, 109, 149, 199, 213, 214

About the Authors

Nita H. Shah received her PhD in Statistics from Gujarat University in 1994. From February 1990 until now Professor Shah has been Head of the Department of Mathematics in Gujarat University, India. She is a postdoctoral visiting research fellow of University of New Brunswick, Canada. Professor Shah's research interests include inventory modeling in supply chains, robotic modeling, mathematical modeling of infectious diseases, image processing, dynamical systems and their applications, etc. She has published 13 monograph, 5 textbooks, and 475+ peer-reviewed research papers. Four edited books have been prepared for IGI-global and Springer with coeditor Dr. Mandeep Mittal. Her papers are published in high-impact journals such as those published by Elsevier, Interscience, and Taylor and Francis. According to Google scholar, the total number of citations is over 3334 and the maximum number of citations for a single paper is over 177. She has guided 28 PhD Students and 15 MPhil students. She has given talks in USA, Singapore, Canada, South Africa, Malaysia, and Indonesia. She was Vice-President of the Operational Research Society of India. She is Vice-President of the Association of Inventory Academia and Practitioner and a council member of the Indian Mathematical Society.

Falguni S. Acharya is a Professor and Head of the Department of Applied Sciences and Humanities, Parul University, Gujarat, India. Dr Acharya has 23 years of teaching experience and 13 years of research experience. Research interests are in the fields of mathematical control theory in differential and fractional differential systems/inclusions with impulse and mathematical modeling of dynamical systems. She has published 16 articles, one international book and one book chapter.